ELECTRICAL INSTALLATION TECHNOLOGY
Third edition

ELECTRICAL INSTALLATION TECHNOLOGY
Third edition

Michael Neidle, FIElecIE, TEng, ASEE (Dipl)

Associate Member of the Institution of Electrical Engineers

Butterworths

London Boston Durban Singapore Sydney Toronto
Wellington

First published 1970
  Reprinted 1973, 1974
Second edition 1975
  Reprinted 1976, 1977, 1978, 1980
Third edition 1982
  Reprinted 1985

British Library Cataloguing in Publication Data

Neidle, Michael
  Electrical installation technology. – 3rd ed.
  1. Electric engineering
  I. Title
  621.3   TK145

  ISBN 0-408-00360 X

Printed in Great Britain by
The Whitefriars Press Ltd, Tonbridge

PREFACE

This book has been written primarily as a textbook for the City and Guilds Electrical Installation Work, Course C (236). It is also designed to assist students taking the Electrical Installation Technicians (285) and Electrical Technicians Courses (281) and will be found to be extremely helpful to students working for the ASEE Diploma. It contains the requisite material suitable for those who wish to qualify for higher positions such as technician, foreman or supervisor; which in turn may lead to the posts of supervising engineer or contracts manager. Where considered necessary slightly more advanced aspects have been added.

This volume seeks to cover the wide range of subjects that come under the headings of electrical science, installations and regulations. To maintain a balance some compression has been found necessary but this has been kept to a minimum. The constant aim has been to combine theory with practice so as to enable the reader to appreciate logically the principles involved.

For this edition a thorough revision has been undertaken to bring all the material up-to-date. In particular to comply with changes in wiring techniques demanded by the 15th Edition of the I.E.E. Wiring Regulations and the new Course C syllabus. The purely technical side has been augmented by a chapter dealing with the important topics which come under the heading of site and office management. Thus the subject matter is also applicable to those who have left their apprentice and student days behind and wish to keep up with modern developments in a changing world.

A feature of the book is the many fully-worked examples. This will prove a valuable aid to that self-study which is so necessary to supplement classwork, and is essential for mastery of the subjects. Conscientious working through and clear setting out of the worked and end-of-chapter exercises should ensure the student a good pass.

Michael Neidle

ACKNOWLEDGEMENTS

The author and publishers acknowledge the help given by the following firms and authorities, in the preparation of this book.

Bang & Olufsen Beocord.
Bastian & Allen Ltd.
B.I.C.C.
Blaco.
British Electrical Transformer Co. Ltd.
Brook Motors Ltd.
City and Guilds of London Institute.
Crompton Parkinson Ltd.
Dubilier Condenser Co. Ltd.
Electrical and Radio Trading.
Electrical Development Association.
Electrical Times.
E.E.P.T.U.
Evershed & Vignoles Ltd.
Falks Ltd.
G.E.C.–A.E.I.
Greenwood Airvac Conduits Ltd.

Heatrae Ltd.
Illuminating Engineering Society.
Institution of Electrical Engineers.
Johnson & Phillips Ltd.
Morelite Engineering Ltd.
MK Electric
Nalder Bros. & Thompson Ltd.
Nife Batteries Ltd.
Northern Counties Technical Examination Committee.
Power Centre Co. Ltd.
Sangamo Weston Ltd.
Sterdy Telephones Ltd.
Union of Lancashire and Cheshire Institutes.
Walsall Conduits Ltd.
Welsh Joint Education Committee.

CONTENTS

CHAPTER 15   ILLUMINATION   295

CHAPTER 16   HEATING   321

Chapter 1

ELECTROMAGNETISM

1.1. Introduction

All electrical machinery and many other electrical devices require magnetic fields for their operation, *these fields being usually set up by electrical means*. In the electric motor, the fields or magnetic fluxes interact to exert mechanical forces, similar in many ways to the forces of attraction and repulsion of freely suspended bar magnets, the purpose of machine design being to make the motor perform continuous rotation efficiently. For generators, conductors are made to rotate by external means and to cut the magnetic fields, thereby causing e.m.f.s to be generated in the conductors; alternatively, stationary conductors may be cut by rotating fields.

The inductive effect set up by varying magnetic fields in windings must also be taken into account by the electrical installation student, for instance, when calculating the circuit loadings of fluorescent fittings or discharge lamps. Inductance also introduces problems of power factor.

Large-scale distribution of electrical energy in Britain would be quite impossible without that wonderful item of electrical equipment known as the transformer. Yet its modest brother, the bell transformer, works on exactly the same principle. Containing no moving parts, its method of operation is almost wholly based on the performance of magnetic fields.

1.2. Domain Theory of Magnetism

According to an earlier theory, the molecules of magnetic substances were themselves considered as tiny magnets each with a north and south pole. If the substances exhibited no external magnetism it was stated to be because the molecules point in random directions so that there is no external effective field. This kind of arrangement is shown in Fig. 1.1(*a*) where each arrowhead depicts a miniature North magnetic pole. Figure 1.1(*b*) shows the material 'saturated', i.e. when all the molecules' magnets point in one direction.

While the general ideas of this molecular theory give a reasonable

1

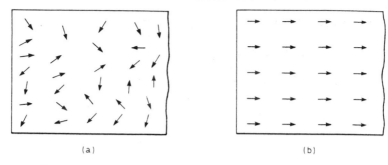

Fig. 1.1. (*a*) Unmagnetised piece of iron bar. (*b*) Fully magnetised.

explanation of magnetic phenomena, it does not explain how the magnetic fields arise in the first place. For such an understanding we have to consider the atom with its revolving electrons. In addition to this orbital motion each electron has a twisting or spinning movement about its axis somewhat on the lines of a spinning top. The gyration or rotational movement of the electron may be likened to a current round a minute circular path or ring and produces a field with a magnetic polarity. This polarity is determined by the direction of the spin and follows the corkscrew rule.

Thus it can be seen that both electrons and magnetic fields constitute an inherent part of all matter. Non-magnetic materials show no external fields because there are equal numbers of clock-

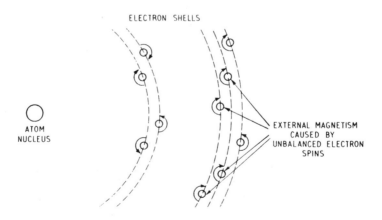

Fig. 1.2. Part of iron atom.

wise and anticlockwise spins, which are, so to speak, self-cancelling, so that the magnetic fields cancel each other out.

Iron is a magnetic material by virtue of the fact that there is a surplus of four electron spins in one direction in each atom. It is this unbalance which produces the external field (Fig. 1.2). Due to its intense speed the orbital movement in three dimensions is often referred to as a shell, which in turn may form sub-shells. Here, in one complete shell, the spins balance, but in the shell which consists of three orbits, there is an unbalance which sets up the external magnetism.

Atoms of the metal cobalt possess an excess of three spins, while nickel gives an excess of two. Atoms of all magnetic materials are grouped together into regions called *domains* with perimeter walls. Within the domain walls, magnetic axes form parallel lines since they have similar directions of magnetism. Thus a true picture can now be obtained if we regard the arrows in Fig. 1.1 as groups of aligned atoms within domain boundaries. Actually magnetisation can be seen as much more involved than the simplified picture as given above, since it results not only from changes in the direction of the domains but is also due to a movement of domain walls making some of the domains increase in size at the expense of others. As an indication of the microscopic sizes involved, the width of a domain approximates 0·1 mm and contains about 10^{15} (1 000 000 000 000 000) atoms.

1.3. Magnetic Materials

Practical magnets may be made by inserting a specimen in a coil through which a heavy d.c. current is passed. The effect of the external field from the coil is to make the domains turn so as to line-up in one direction.

Magnetic materials are often called ferro-magnetic and usually consist of iron, steel or magnetic alloys. Modern magnets often combine the iron with nickel, cobalt, copper or tungsten in order to obtain superior characteristics.

1.4. Magnetic Field Properties

Electrical engineers are largely concerned with the external field which may be defined simply as the space where the magnetic influence is felt. Since the field is invisible it is normal to represent the field by 'lines of force', these being also termed lines of magnetic flux or flux lines.

Although lines of force have no real existence they are a very useful concept as the strength, or *density*, of a field can be given by the number of lines per unit area. It is also universally accepted that the *direction of a magnetic field at any point is shown by the direction of the north pole of a compass needle when placed at that position.*

1.5. The Magnetic Circuit

We should also bear in mind that lines of force often consist of closed loops to form magnetic circuits. In a bar magnet the external path of the magnetic circuit may be plotted by a compass needle, the loop being completed within the bar magnet (Fig. 1.3). The ordinary electric trembler bell, for its action, requires the armature to complete quite a strong magnetic circuit. This may be seen in Fig. 1.4 where the coils are omitted for the sake of clarity.

1.6. Magnetic Behaviour

The following guides, which have been obtained by practice, are extremely useful as a help to understanding the way in which magnetic fields and electromagnetic apparatus behave.

(1) *Lines of force never cross,* therefore if a compass needle is placed in a position where it is under the influence of two magnetic fields, the needle will point in the direction of the *resultant* field. The actual direction will in fact be determined by which field is stronger. Figure 1.3 shows that at each position of the compass needle, the

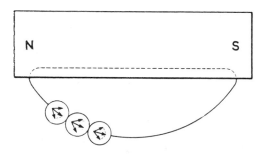

Fig. 1.3. A simple magnetic circuit. The tracing out of lines of force by a compass needle also illustrates the principle that lines of force never cross. The needle takes up a *resultant* position dependent upon the vector forces exerted on it from the N and S poles.

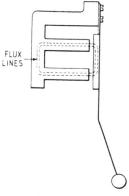

Fig. 1.4. Magnetic circuit of electric bell.

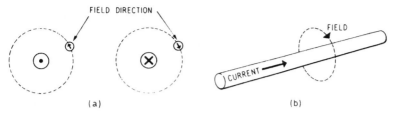

(a)

(b)

CORKSCREW RULE. Field direction around a conductor carrying a current, as shown by a compass needle.

Reversing the current in the conductor also reverses the field direction.

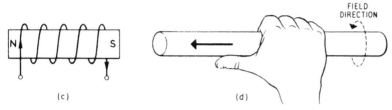

(c)

(d)

SOLENOID RULE. Indicates that a south pole is produced at the end of the coil where the current is flowing in a clockwise direction. Current flowing in an anticlockwise direction produces a north pole.

GRIP RULE. If the right hand grips a conductor with the thumb pointing in the direction of the current, then the fingers point in the direction of the field. If the grip rule is applied to a solenoid, as (c), the thumb points in the direction of the north pole when the fingers point in the direction of the current.

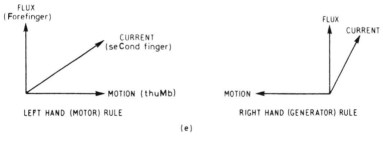

LEFT HAND (MOTOR) RULE

RIGHT HAND (GENERATOR) RULE

(e)

FLEMING RULE. Relating directions of current, flux and motion in motors and generators.

Fig. 1.5. Magnetic field directions.

force of repulsion and attraction *vectorially* combine to produce the resultant position.

(2) Lines of force act as if they are in tension, i.e. as stretched elastic lines.

(3) Parallel lines of flux which are in the same direction tend to repel one another.

(4) A current-carrying conductor if free to move, and placed in a magnetic field, will travel from the stronger part of the field to the weaker.

Since current, when it flows through a wire, produces a definite and invariable magnetic field direction, rules for showing the relationship between the current and magnetic directions can be formulated. These rules together with some mnemonics, or memory aids, are illustrated in Fig. 1.5.

1.7. Flat-hand Rule

The phrase '*Motors* travel on the *left*' may assist in remembering the left-hand rule. The motor and generator rules were formulated by Professor Fleming from the concept that directions of current, flux and motion for current-carrying conductors are mutually at

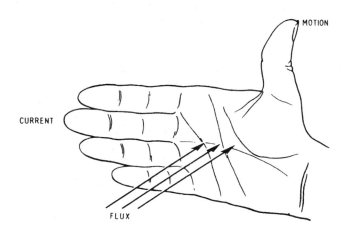

Fig. 1.6. Flat-hand rule. Most students will find it simpler to use than the Fleming Rule. The right hand applies to generators and the left hand to motors.

right angles. Unfortunately, the application of these rules requires some finger dexterity. For this reason the author prefers the use of the 'flat-hand rule' (Fig. 1.6).

1.8. Units and Calculations

In the S.I. system, lengths are measured in metres (m), areas in square metres (m^2) and mass in kilograms (kg), while seconds (s) are used for time. The newton (N) is taken as the unit of force and is defined as the force necessary to give a mass of 1 kg an acceleration of one metre per second per second (1 m/s^2).

The unit of torque or twisting moment is the newton metre (N m) being the torque due to the force of 1 newton acting at 1 metre radius.

1.9. Magnetic Flux (Symbol Φ)

The unit of magnetic flux is the weber (Wb) and may be taken as equal to one line of force or flux-line. Sub-units are often used: 1 milliweber (mWb) is equal to 10^{-3} Wb and a microweber (μWb) equals 10^{-6} Wb.

1.10. Flux Density (Symbol B)

Flux density is measured in teslas (T). The flux density is a measure of the field strength since it indicates the density of the field or the number of lines in each square metre of the field.

1.11. Magneto-motive Force or m.m.f. (Symbol F)

The unit of m.m.f. is the ampere-turn (At or NI), but since the number of turns is simply a multiplier, A is often accepted as the unit for m.m.f. The field set up by a coil is obtained by a current flowing through the winding. The effect of additional turns, with the same current, is to increase the flux unless saturation has been reached.

1.12. Magnetising Force (Symbol H)

The magnetising force must not be confused with m.m.f. Its unit is the ampere-turn/metre *length* of magnetic path (At/m). In symbolic form, $H = NI/l$ where l is the length of the magnetic path in metres.

1.13. Permeability of Free Space, i.e. Air (Symbol μ_0)

Permeability may be regarded as the ability to allow a magnetic field to be set up. It is in fact a *ratio* of flux density to magnetising force and is given as a fixed number. The permeability of free space (i.e. air) is given by

$$\mu_0 = \frac{B}{H} = 4\pi \times 10^{-7} \text{ H/m}$$

The unit henry per metre has been added to comply with the SI system.

1.14. Relative Permeability (Symbol μ_r)

Relative permeability always occurs where iron, or any other magnetic material, is present in a magnetic circuit. μ_r is given by a number, without any units and has a value dependent upon the particular grade of iron and, as we shall see later, the particular point on the *B–H* curve. *Relative permeability multiplied by permeability of free space gives absolute permeability.* Thus for a magnetic material

$$\mu_0\mu_r = \frac{B}{H}$$

Example 1.1

An iron ring has a cross-sectional area of 400 mm² and a mean diameter of 25 cm. It is tightly wound with 500 turns. If the value of the relative permeability is 250, find the total flux set up in the ring. The coil resistance is 474 ohms and the supply voltage equals 240 V.

The conditions of the problem may be represented by Fig. 1.7. Current through coil,

$$I = \frac{V}{R} = \frac{240}{474} \qquad = 0{\cdot}506 \text{ A}$$

Mean length of magnetic path,

$$l = \pi \times \text{diameter}$$

$$= \pi \times \frac{25}{10^2} \qquad = 0{\cdot}7854 \text{ m}$$

Magnetising force,

$$H = \frac{NI}{l} \quad \text{where } N \text{ is number of turns and}$$

$$I \text{ is current}$$

$$= \frac{500 \times 0{\cdot}506}{0{\cdot}7854} \qquad = 322{\cdot}3 \text{ A/m}$$

Now $$\frac{B}{H} = \mu_0\mu_r$$

$$\therefore \qquad B = H\mu_0\mu_r$$
$$= 322\cdot3 \times 4\pi \times 10^{-7} \times 250 = 0\cdot1014 \text{ T}$$

But $$B = \frac{\Phi}{a}$$

$$\therefore \qquad \Phi = B \times a$$

$$= 0\cdot1014 \times \frac{400}{10^6}$$

(note division by 10^6 to convert from mm^2 to m^2)

$$= 0\cdot000\ 040\ 6 \text{ Wb} \qquad = 40\cdot6\ \mu\text{ Wb}$$

1.15. Leakage and Fringing

The above closely-wound solid-iron ring, often called a *toroid*, is the simplest example of a magnetic circuit and is the easiest to calculate as there is no magnetic leakage or fringing (Fig. 1.8). The type of question given above in Example 1.1 often includes a part which requires the student to answer the question 'what would be the effect of a saw-cut in the ring?'. The saw-cut produces a certain amount of fringing and therefore causes a reduction in the ring flux; alternatively, more ampere-turns would be required to set up the same flux

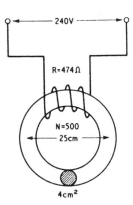

Fig. 1.7 (left). Simple magnetic problem (Example 1.1). Find the total flux set up in the ring.

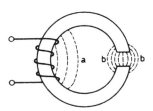

Fig. 1.8 (above). Leakage and fringing. Leakage lines at *a*. Fringing is represented by the bulging lines at *bb*, thus increasing the cross-sectional area of the gap, which in turn reduces the flux density.

in the ring. Example 1.2 shows how the additional ampere-turns can be calculated.

1.16. Magnetic Ohm's Law

With certain reservations there is quite a good analogy between electric and magnetic circuits:

ELECTRIC	MAGNETIC
The e.m.f. forces a current against a resistance.	The m.m.f. produces a flux in opposition to the magnetic reluctance.

$$\text{Current} = \frac{\text{Electromotive-force}}{\text{Resistance}} \qquad \text{Flux} = \frac{\text{Magneto-motive force}}{\text{Reluctance}}$$

$$I = \frac{E}{R} \qquad\qquad \Phi^* = \frac{F}{S}$$

With units as already given

$$\text{Reluctance} = \frac{l}{\mu_0 \mu_r a}$$

Reluctance may be defined as the property of a magnetic circuit (or part of a magnetic circuit) which opposes the passage of a magnetic flux through it.

The student will note that there is thus an analogy between reluctance and resistance, where $R = \dfrac{\rho l}{a}$, ρ being the symbol for resistivity.

The method of employing $\Phi = \dfrac{F}{S}$ is very useful in solving problems involving magnetic circuits. It could be used to solve the previous example, but the particular application is for involved magnetic circuits consisting of different cross-sectional areas and/or different materials.

* Proof

$$\mu_0 \mu_r = \frac{B}{H} = \frac{\Phi/a}{NI/l} = \frac{\Phi l}{NIa}$$

∴

$$\Phi l = \mu_0 \mu_r NIa$$

or

$$\Phi = \frac{NI}{l/\mu_0 \mu_r a} = \frac{F}{S}$$

1.17. Magnetic Series Circuit

Practical magnetic circuits require the determination of the number of ampere-turns to produce a particular flux. When the magnetic paths are in *series* the total ampere-turns may be calculated as follows:

$$\text{Magnetic flux } (\Phi) = \frac{\text{m.m.f. (ampere-turns)}}{S \text{ (reluctance)}}$$

∴ $$\text{Total ampere-turns} = \Phi S_1 + \Phi S_2 + \ldots$$

where Φ is the flux required and S_1, S_2 etc. are equal to the reluctances of the various parts.

Example 1.2

An iron ring of C.S.A. 600 mm² is closely wound with insulated wire and has a saw-cut of 2 mm. Calculate the total ampere-turns to produce a flux of 0·1 mWb if the mean length of the magnetic path is 300 mm and the relative permeability of the iron is 470.

The student will usually find it is worthwhile to put the information contained in these types of questions into the form of a suitable diagram (Fig. 1.9).

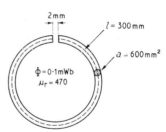

Fig. 1.9. Saw-cut in iron ring (Example 1.2).

$$\text{Flux} = \frac{\text{m.m.f.}}{\text{reluctance}}$$

∴ $$\Phi = \frac{NI}{S}$$

and $$\text{Ampere-turns} = \Phi \times S$$

Referring to the iron ring

$$S = \frac{l}{\mu_0 \mu_r a}$$

Remembering to convert mm to metres and mm² to m²:

$$S = \frac{300 \times 10^6 \times 10^7}{4\pi \times 470 \times 600 \times 10^3} = 850\,000$$

$$\text{Ampere-turns} = \Phi S$$

$$= 0\cdot1 \times 10^{-3} \times 850\,000 = 85$$

Referring to the air-gap,

$$S = \frac{l}{\mu_0 a}$$

$$= \frac{2 \times 10^7 \times 10^6}{4\pi \times 10^3 \times 600} = 2\,650\,000$$

$$\text{Ampere-turns} = 0\cdot1 \times 10^{-3} \times 2\,650\,000 = 265$$

$$\therefore \quad \text{Total ampere-turns} = 85 + 265 = \underline{350}$$

The solution demonstrates the superiority of the iron as a flux 'conductor'. Many more ampere-turns are required to produce the magnetic flux through 2 mm of the air-gap than through the metal.

1.18. Force on a Conductor

In the simple case of a current-carrying conductor placed at right angles to a magnetic field, the conductor will experience a mechanical force in a direction as given by either the Fleming or the flat-hand rule (Figs. 1.5 and 1.6). It is clear that this force must be proportional to

Strength of magnetic field (B)
Magnitude of current (I)
Length of conductor (l)

i.e.   Force $\propto BIl$.

Force $= BIl$ newtons where B is in teslas
I in amperes
l in metres

(A memory aid: Force $= BIl(l)$ Newton.)

We are now in a position to appreciate an alternative definition of flux density. A flux density of 1 tesla is produced when a conductor carrying 1 A, when at right angles to a magnetic field, experiences a force of 1 newton on every metre of conductor length.

Example 1.3

(i) *A single armature conductor lying at right angles to the magnetic field of a 2-pole d.c. motor carries a direct current. Make a sketch or*

Fig. 1.10. Motion of a current-carrying conductor in a magnetic field
(Example 1.3). (*a*) Single conductor, (*b*) looped conductor.

*diagram of the above arrangement, choosing directions of current and
magnetic field and show the direction in which the conductor will tend
to move.*

*From the above show how the armature of a 2-pole d.c. motor will
rotate under the influence of a magnetic field and armature current.*

(ii) A single armature conductor as above is 30 *cm long and the flux
density of the magnetic field is* 1·4 *T. Calculate the force on the
conductor when it carries* 94 *A.*   [C & G]

(*i*) Use of the flat left-hand or the Fleming left-hand rule applying
to motors (Section 1.7) gives the direction in which the single
conductor will tend to move if positioned as at Fig. 1.10(*a*). If the
conductor is in the form of a loop (Fig. 1.10(*b*)) it can now be seen to
form part of an armature and, with current and field directions as
shown, the loop would tend to rotate in an anticlockwise direction.

(*ii*) $F = BIl$ newtons

$$= 1·4 \times 94 \times 30 \times 10^{-2} \qquad = 39·48 \text{ newtons}$$

It is instructive to employ the superimposition of magnetic fields
as an alternative method for obtaining the direction of motion
(Fig. 1.11).

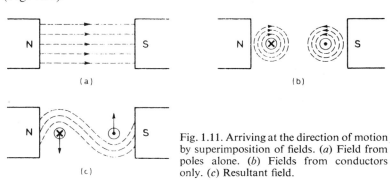

Fig. 1.11. Arriving at the direction of motion
by superimposition of fields. (*a*) Field from
poles alone. (*b*) Fields from conductors
only. (*c*) Resultant field.

The fields are strengthened above the left-hand conductor and below the conductor on the right, since at these positions the fields point in the same direction. Similarly weakening of the fields occur under the left-hand and above the right-hand conductors. One of the results of the conductor-magnetic field relationship is that the current-carrying conductors tend to move out of a strong field into a weak field. The resultant field may also be seen as to consist of stretched elastic bands which, as they straighten out, tend to rotate the conductor loop in an anticlockwise direction.

Example 1.4

A long straight single conductor connected to the live pole carries a d.c. current of 500 A. The neutral conductor is placed at a sufficient distance away so that its magnetic field does not affect that of the live conductor. Find the magnetising force and flux density at a distance of 50 mm from the conductors.

The live and neutral conductors may be considered as forming *one* complete turn.

$$\therefore \qquad \text{m.m.f.} = NI$$

$$= 1 \times 500 \qquad = 500 \text{ ampere-turns}$$

Considering the magnetic circuit, length of path is that of the circumference of a circle of radius 50 mm (see Fig. 1.12),

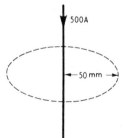

500A

←— 50 mm —→

Fig. 1.12. Magnetic circuit at 50 mm from a straight conductor (Example 1.4).

$$l = 2\pi r = 2\pi \times 50 \times 10^{-3} \text{ metres} = 0\cdot3142 \text{ m}$$

$$\text{magnetising force,} \quad H = \frac{NI}{l}$$

$$= \frac{500}{0\cdot3142} = 1590 \text{ A/m}$$

As no iron is present $\dfrac{B}{H} = \mu_0$

$$B = \mu_0 H$$
$$= 4\pi \times 10^{-7} \times 1590 = 2 \text{ mT (millitesla)}$$

1.19. Force between Parallel Conductors

Figure 1.13 illustrates the plan view of busbars forming rising mains. In between the live and neutral conductors of Fig. 1.13(a), the parallel lines of flux, as they point in the same direction, tend to repel each other, resulting in a mechanical force of repulsion

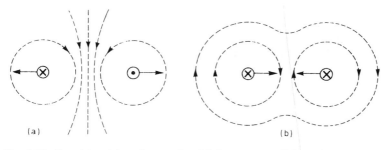

(a)   (b)

Fig. 1.13. Repulsion (a) and attraction (b) between parallel conductors due to magnetic fields set up by currents carried by the conductors.

between the busbars. Figure 1.13(b) represents the resultant field due to two phase conductors adjacent to each other; here, the fields between the conductors cancel each other out. The external resultant field, behaving as stretched elastic lines, tends to pull the busbars towards each other. Alternatively, the conductors can be viewed as moving away from the relatively strong external field into the weaker field between conductors.

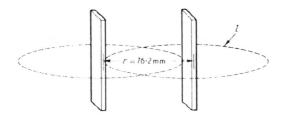

$r = 76\cdot2\,\text{mm}$

Fig. 1.14. Example 1.5. Magnetic circuits from two busbars:
l = length of magnetic path.

Example 1.5

Two busbars, each 20 metres long, feed a circuit and are spaced apart at a distance of 80 mm in between centres. If a short-circuit current of 20 000 A flows through the conductors, calculate the force per metre between the bars.

Considering each busbar acting as a single turn:

$$\text{Ampere-turns} = 20\,000 \times 1$$

From Fig. 1.14,

Length of magnetic circuit $(l) = 2\pi r$

$$= 2\pi \times 80 \times 10^{-3} \text{ metres} = 0\cdot5027 \text{ m}$$

magnetising force $(H) = \dfrac{NI}{l}$

$$= \frac{20\,000}{0\cdot5027} = 39\,785 \text{ At/m}$$

∴ Flux density (B) at 80 mm radius

$$= \mu_0 H$$
$$= 4\pi \times 10^{-7} \times 39\,785 = 0\cdot05 \text{ T}$$

∴ Total force on each conductor from each other for a conductor length (L) of 20 m

$$= BIL \text{ newtons}$$
$$= 0\cdot05 \times 20\,000 \times 20 = 20\,000 \text{ N}$$

∴ Force on each metre of busbar

$$= \frac{20\,000}{200} = \underline{1000 \text{ N}}$$

It should be noted that the force will be that of repulsion.

The result of force produced by a heavy short-circuit can be seen in the photograph (Fig. 1.15).

The force acting between two parallel conductors has led to the modern definition of the ampere. *One ampere is that current flowing between two long parallel conductors spaced* 1 *metre apart which produces a force between them of* 2×10^{-7} *N/m length of run* (*Fig.* 1.16). Strictly speaking the conductors should be infinitely long and spaced in a vacuum, but the above definition is normally taken as sufficient for practical purposes.

Fig. 1.15. Condition of busbars after being subjected to short-circuit current considerably in excess of their rating.

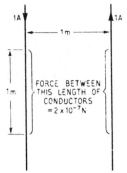

Fig. 1.16. Illustration of the electromagnetic definition of the ampere.

1.20. *B–H* Curves

While certain magnetic calculations can be developed from first principles, practical working is normally based on performance graphs as supplied by manufacturers of magnetic materials. These are known as *B–H* magnetisation curves, a basic pattern of which is illustrated in Fig. 1.17.

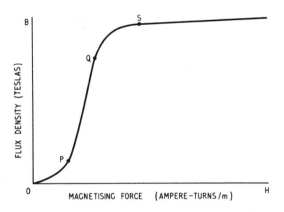

Fig. 1.17. The basic *B–H* curve.

The interpretation of the curves in terms of the domain theory (1.2) would show no alignment at zero position and maximum alignment of the magnetic domains at saturation, where the curve flattens out from S onwards.

The curve may also be seen to consist of three distinct parts. From O to P, the magnetising force is just sufficient to rotate the atomic magnetic groupings within the iron. The portion consisting of P to Q is usually considered as the working portion. Since here the slope is steepest, the maximum flux density (B) can be produced for a given magnetising force (H). In addition the permeability, or the ratio B/H, has a constant value over this range.

At S, saturation begins to be reached and it would be uneconomical to work the material beyond this point. It can be clearly observed that here any increase in magnetising force would do little to produce a stronger field. Magnetising force NI/l is obviously proportional to current, therefore for the sake of economy the magnetising force must be kept to a minimum for any particular value of flux density.

Typical working curves may be seen in Fig. 1.18. Silicon steel is widely used for transformer laminations, as well as for motors and transformers. The use of cast-iron is almost wholly restricted to the yokes and pole pieces of small d.c. machines. The superior performance of cast-steel over cast-iron makes for reduced dimensions.

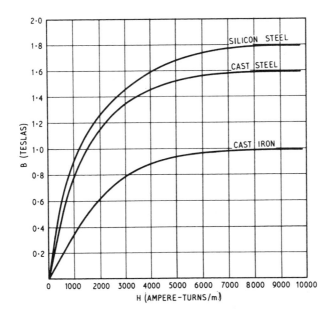

Fig. 1.18. Typical manufacturer's magnetism curves.

Grain-oriented magnetic silicon iron has lower losses and permits higher working flux densities than with normal hot-rolled iron. By a special process, the iron crystals are set in one direction ('grain-oriented'), resulting in increased permeability. This makes for more economical working in transformers and fluorescent chokes.

Silicon steel, cast-steel and cast-iron are examples of 'soft' magnetic metals. These materials gain and lose their magnetism easily, possessing a high permeability (μ_r being approximately equal to 5000) and low residual magnetism. Permanent magnets, which are now increasing in everyday use, are of the 'hard' type with a natural high residual magnetism. The modern strong magnets such as Alnico and Alcomax, as their names suggest, are made from magnetic alloys.

Example 1.6

(a) *Referring to Fig.* 1.18, *what value of flux density would be produced by a magnetising force of* 6500 *ampere-turns/metre in silicon steel, cast steel and cast iron respectively?*

(b) *If the average length of flux path is* 0·4 *m and the number of turns* = 200, *calculate the magnetising current required.*

(a) For a value of 6500 ampere-turns/metre, the flux densities are:

	T
Silicon steel	1·76
Cast-steel	1·56
Cast-iron	0·98

(b) $$H = \frac{NI}{l}$$

∴ $$I = \frac{Hl}{N} = \frac{6500 \times 0\cdot4}{200} \qquad = \underline{13\ A}$$

A further example brings out the usefulness of the *B–H* curve method in avoiding the necessity for calculating reluctance.

Example 1.7

A ring made of cast steel has a mean circumference of 500 *mm and a cross-sectional area of* 0·05 m^2. *It is closely wound with a coil consisting of* 200 *turns. What current is required to produce a flux of* 0·05 *Wb in the ring? (Refer to Fig.* 1.18.)

$$\text{Flux density} = \frac{\Phi}{a} = \frac{0\cdot05}{0\cdot05} = 1\ T$$

This corresponds to a value of magnetising force (H) of 1520 ampere-turns/m (Fig. 1.18).

But
$$H = \frac{NI}{l}$$

$\therefore$
$$I = \frac{Hl}{N} = \frac{1520 \times 0 \cdot 5}{200} \qquad = \underline{3 \cdot 8 \text{ A}}$$

1.21. The Hysteresis Loop

The complete *cycle of magnetisation* is given by this loop (Fig. 1.19). The word 'hysteresis' is derived from the Greek and means lagging behind. By magnetising the material up to saturation, the curve at first follows the normal *B–H* curve (0a). If the magnetising force is reversed by reversing the current, the shape of the curve does not mirror the original and is shown by ac. The lagging-behind effect can be seen at point b, where H has been reduced to zero but B has still a value 0b often called remanence or remanent flux density. To destroy this residual magnetism a further quantity of H, 0c coercive force, is required.

The complete loop really amounts to subjecting the material to an alternating magnetism force. Thus on alternating current the com-

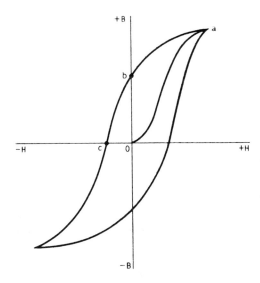

Fig. 1.19. The hysteresis loop.

plete. path is traced out during each cycle. The area enclosed by the loop represents a loss of energy. This is accounted for, on the domain theory, by the energy required first to 'line-up' the domains which then reset themselves in a random pattern as demagnetisation is reached. The process which is continuous is sometimes referred to as molecular friction.

EXERCISES

1. A conductor is suspended between the poles of an electromagnet so that it is at right angles to the lines of magnetic flux. Explain, *chiefly with the aid of sketches*, the effect on the conductor of

(*a*) passing a current through it,

(*b*) passing a higher value of current through it,

(*c*) passing a higher value of current through the coils of the electromagnet,

(*d*) reducing the gap between the poles of the electromagnet by bringing the poles closer to each other.

In (*a*) above what is the effect

(*i*) if the direction of current flowing through the conductor is reversed and

(*ii*) if the gap between the poles of the electromagnet is reduced by attaching strips of brass to each pole piece?   [T]

2. (*i*) What is the flux density produced by a magnetising force of 500 000 ampere-turns/m in air?

(*ii*) A straight conductor when carrying a current of 55 A is positioned at right angles to a field with a flux density of 2·4 T. Calculate the force on the conductor if it is 1 m long.

3. A pair of rising mains has a spacing of 200 mm between centres. If each conductor carries 500 A determine the force between the conductors for each 10 m length of run.

4. State what you understand by the following terms and give units where applicable: (*a*) magnetising force; (*b*) magnetomotive force, (*c*) permeability, (*d*) hysteresis.

5. (*i*) Describe an experiment which illustrates both the attractive and repulsive forces set up by magnetic fields. (*ii*) Name six magnetic materials. (*iii*) Sketch the magnetic circuit of an item of electrical equipment.

6. Draw an analogy between a series electric circuit and a series magnetic circuit. From this develop an expression for flux in terms of m.m.f. and reluctance.

An iron ring has a cross-sectional area of 400 mm^2 and a mean diameter of 250 mm. An air-gap of 1 mm has been made by a saw-cut across the section of the ring. If a magnetic flux of 0·3 mWb is required in the air-gap, find the current necessary to produce this flux when a coil of 400 turns is wound on the ring. The iron has a relative permeability of 500.

7. Sketch a curve showing the relationship between the flux density and the magnetising force for a solenoid having (a) an air core and (b) a ring iron core. State the units in which the two quantities are measured and define the permeability of free space.

8. Give *three* examples of the practical use of the magnetic effect of an electric current.

With the aid of diagrams show the magnetic field surrounding two parallel wires carrying conventional current.

(*a*) in the same direction

(*b*) in opposite directions

indicating the directions of the currents and the lines of force.

State, for each case, whether the force will tend to bring the wires nearer together or to separate them.

State a rule for determining the direction of the force between the wires.   [T]

9. (*a*) Explain what is meant by 'hysteresis' in a magnetic material. (*b*) Sketch typical hysteresis curves for a material (i) with high hysteresis loss and (ii) with low hysteresis loss. (*c*) Mark on curve (*b*)(i) the saturation point, the remanent flux density and the coercive force. (d) What effect will frequency have on the power loss?   [C]

Chapter 2

INDUCTANCE

2.1. Inductance

When a varying current passes through a coil it sets up a varying magnetic field which has the effect of inducing an e.m.f. in any circuit linking with the flux. It is the change in flux-linkages (Fig. 2.1) which causes an e.m.f. to be set up. Some students understand this effect more easily by visualising the conductors as being 'cut' by the field when it collapses or is being created.

Inductance is due to magnetic linkages. Any apparatus which involves a large number of turns and especially if linked by a magnetic circuit will possess a high inductance. The symbol for inductance is L and the unit is the henry (H). A coil is said to possess *an inductance of 1 H when current changing at the rate of 1 ampere/ second (A/s) induces an e.m.f. of 1 V.*

2.2. Induced E.M.F.

The induced e.m.f. referred to above is in the nature of a back e.m.f.; LENZ'S LAW gives the direction of the induced e.m.f. as being such *as to oppose the change in current producing this e.m.f.* Inductance causes a delay in the current reaching its final value and prevents the instantaneous collapse of the current. While Fig. 2.2 shows the general shape of the curves for a coil containing resistance and

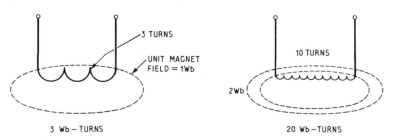

Fig. 2.1. Flux-linkages expressed in weber-turns.

24

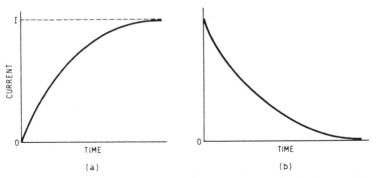

Fig. 2.2. Effect of inductance on current, (a) at instant of switching on, (b) at instant of switching off.

inductance, they are derived more exactly by means of the time constant (T) of coil:

$$T(\text{s}) = \frac{L(\text{henrys})}{R(\Omega)}$$

and is defined as the time taken for the current to reach its final value, should the initial rate of current change (A/s) be maintained.

Example 2.1

A coil of resistance 8 Ω and inductance 2 H is connected to a 20-V supply.
    (a) Find: (i) the final steady value of the current: (ii) the time constant of the coil; (iii) the initial rate of change of current in the coil; (iv) the energy stored in the coil when the current reaches its final value.
(b) Sketch a graph showing the current from 0 to 1 s.   [C]

    (a)(i) Final steady current $= \dfrac{V}{R}$

$$= \frac{20}{8} \qquad\qquad = \underline{2.5\ \text{A}}$$

    (ii) Time constant $= \dfrac{L}{R}$

$$= \frac{2}{8} \qquad\qquad = \underline{0.25\ \text{s}}$$

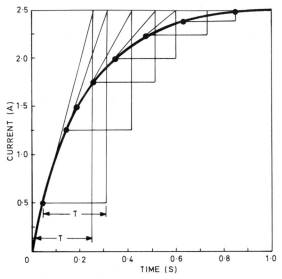

Fig. 2.3. Curve of current growth in an inductive coil.

(iii) Initial rate of change of current $= \dfrac{I}{t}$

$$= \frac{2 \cdot 5}{0 \cdot 25} \qquad\qquad = \underline{10 \text{ A/s}}$$

(iv) Energy stored in coil $= \frac{1}{2}LI^2$

$$= \tfrac{1}{2} \times 2 \times 2 \cdot 5^2 \qquad\qquad = \underline{6 \cdot 25 \text{ J}}$$

(b) The derivation of the exponential curve as shown in Fig. 2·3, by making use of the time constant, is self-explanatory.

A graphical solution to switching *off* an inductive circuit follows the same method as employed for Example 2.1, except that the curve is reversed (Fig. 2.2(b)).

The inductance in henrys may also be expressed by the number of *flux-linkages set up per ampere*.

Suppose I amperes through N turns (NI flux-linkages) produce a flux of Φ webers and the current grows from its initial value 0 to its final value I in t seconds, then average rate of change of current (di/dt in the calculus notation)

$$= \frac{\text{change in current}}{\text{change in time}} = \frac{I}{t} \text{ amperes/second}$$

But induced e.m.f., e volts

$= $ inductance (henrys) $\times$ rate of change of current (A/s)

$= -L\dfrac{\mathrm{d}i}{\mathrm{d}t}$ volts (the minus sign denotes a back e.m.f.)

Since an inductance of 1 H is set up when current change at this rate induces 1 V,

$\therefore$ Inductance $\quad L = \dfrac{\text{induced e.m.f.}}{\text{rate of current change}} = \dfrac{e}{I/t}$

but $\qquad e = $ rate of change of flux-linkages $= \dfrac{\Phi N}{t}$

$\therefore \qquad\qquad L = \dfrac{\Phi N/t}{I/t}$

$\qquad\qquad\quad = \dfrac{\Phi N}{I} = $ flux-linkages/ampere

Example 2.2

The e.m.f. induced in a coil is 100 V when the current through it changes from 1 to 10 A in 0·1 second. Calculate the inductance of the coil.

Average rate of change of current $= \dfrac{\text{final current} - \text{initial current}}{\text{time taken}}$

$$= \dfrac{10 - 1}{0·1} = 90 \text{ A/s}$$

Hence $\qquad\qquad L = \dfrac{e}{I/t} = \dfrac{100}{90} \quad = 1·11 \text{ H}$

Example 2.3

The current through a certain coil of inductance 0·4 H is reduced from 10 A to 2 A in 0·06 seconds. Find the average value of the e.m.f. induced in the coil.

Average rate of current increase

$$= \dfrac{\text{final} - \text{initial currents}}{\text{time taken}} = \dfrac{2 - 10}{0·06} = -133·3 \text{ A/s}$$

$$\text{Average e.m.f.} = -L\frac{di}{dt} \text{ volts} = -0\cdot4 \times (-133\cdot3) = \underline{53\cdot32 \text{ V}}$$

By Lenz's law the induced e.m.f. acts in the same direction in this case as the applied e.m.f., i.e. in a direction to prevent the reduction of the current. When current is switched off in a highly inductive circuit, the result will be a momentary arc, with consequent heat rise at switch contacts. The collapse of the current and resultant field induces the back e.m.f. which attempts to bridge across the switch. The electrician may have noticed quite big sparking at the make-and-break contact of an electric bell even when operated off a $4\frac{1}{2}$ V battery.

2.3. The Field-discharge Switch

Example 2.4

The field winding of a machine has an inductance of $2\cdot5$ H and carries a current of 20 A, calculate the induced e.m.f. if the circuit is broken in $0\cdot01$ second.

Since the current is reduced from 20 A to zero in $0\cdot01$ second, the induced e.m.f.,

$$e = -L \times \text{rate of change of current}$$

$$= -2\cdot5 \times \frac{0-20}{0\cdot01} = \underline{5000 \text{ V}}$$

This high value of induced e.m.f., probably many more times than the supply voltage, may damage the insulation of the windings.

A field-discharge switch (Fig. 2.4) may be employed to suppress heavy arcs caused by inductive circuits. It should be noted that the

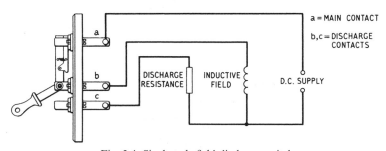

Fig. 2.4. Single-pole field-discharge switch.

contact is made with switch contact at c immediately before the switch is opened at a. The discharge resistance is in parallel with the inductive field and, by forming a local circuit, will be able to absorb the energy of the arc. The resistance, if placed continuously in the circuit would be wasteful. The special discharge switch only permits the resistance to be in parallel across the field during the breaking period of the circuit. The actual value of the discharge resistance may be calculated as shown by Example 2.5.

2.4. Energy Stored in the Magnetic Field

As we have seen, at the moment of switching off the current tends to be maintained. The reason for this is based on the idea that energy is being stored in the magnetic field while the current is increasing. The energy is not used up but is returned to the circuit as the field collapses when the current is switched off. An expression for the energy stored, in joules, can be derived as follows:

Let the current increase from zero to I amperes in t seconds through an air-cored inductance coil (Fig. 2.5) of value L henrys; numerically, the average e.m.f. induced

$$e = \text{inductance} \times \text{rate of change of current}$$

$$= L\frac{di}{dt}$$

but average current increase in air-cored inductor

$$= \frac{I-0}{2}$$

$$= \tfrac{1}{2}I$$

$$\text{Average power} = L \times \frac{I}{t} \times \frac{I}{2} \text{ watts}$$

$$= L\frac{I^2}{2t} \text{ watts}$$

$$\text{Energy} = \text{power} \times \text{time}$$

$$= \frac{LI^2}{2t} \times t \text{ joules}$$

$$= \tfrac{1}{2}LI^2 \text{ joules}$$

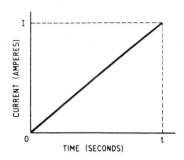

Fig. 2.5. Current increase in air-cored inductor.

Example 2.5

A 6-pole 500-V d.c. shunt generator has a flux/pole of 50 mWb produced by a field current of 10 A. Each pole is wound with 600 turns.

The resistance of the entire field is 50 Ω. If the field is broken in 0·02 second, calculate

   (a) *the inductance of the field coils,*

   (b) *the induced e.m.f.,*

   (c) *the value of discharge resistance so that the induced e.m.f. should not exceed 1000 V.*

(a) Inductance = flux-linkages/ampere = $\dfrac{\Phi N}{I}$

Total inductance for six poles

$$= \frac{6 \times 50 \times 600}{10 \times 10^3} = \underline{18 \text{ H}}$$

(b) Induced e.m.f.

$$= \text{change in flux-linkages/second}$$

$$= \frac{\Phi N}{t} = \frac{50 \times 600}{10^3 \times 0 \cdot 02} = \underline{1500 \text{ V}}$$

(c) At the instant when the field is disconnected from the supply, the field current of 10 A will flow through the local circuit consisting of the field coils and the discharge resistance.

If the induced e.m.f. is not to exceed 1000 V,

$$\text{Total resistance} = \frac{1000 \text{ V}}{10 \text{ A}} = 100 \text{ }\Omega$$

Hence,   Value of discharge resistance = $100 - 50 = \underline{50 \text{ }\Omega}$

Example 2.6

Calculate the energy stored (a) in joules (b) in watt-hours in the magnetic field of a coil when carrying 100 *A, the inductance of the coil being* 0·2 *H.*

(a) $\qquad$ Energy stored $= \frac{1}{2}LI^2$ joules

$$= \frac{1}{2} \times 0 \cdot 2 \times 100^2 = \underline{1000 \text{ J}}$$

(b) $\qquad$ Energy stored $= \dfrac{1000}{60 \times 60} = \underline{0 \cdot 278 \text{ watt-hour}}$

2.5. Inductance in A.C. Circuits—Inductive Reactance

Alternating currents produce a continual current variation. Where applied to an inductor the effect is to induce a back e.m.f. which *continuously opposes the current change.* If the inductance in henrys is multiplied by $2\pi f$ (frequency in Hz), the result is known as INDUCTIVE REACTANCE (symbol X_L) and is measured in ohms. Readers should however be careful to note that reactance is in the nature of a back e.m.f.

2.6. Derivation of $X_L = 2\pi f L \ \Omega$

In a.c. circuits, if frequency $= f$ Hz, then time for 1 cycle $= 1/f$ second. Referring to Fig. 2.6, the maximum value of a.c. is reached in a quarter of a cycle,

∴ $\quad$ Time taken to reach maximum value

$$= \frac{1}{4f} \text{ seconds}$$

Now

$$\text{Rate of current increase} = \frac{\text{current increase}}{\text{time taken}}$$

∴ $\quad$ Rate of current increase during $\frac{1}{4}$ cycle

$$= \frac{I_m}{1/4f}$$

$$= 4f I_m \text{ amperes/second}$$

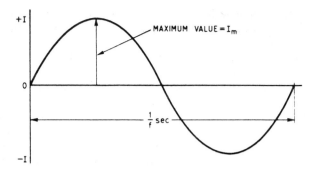

Fig. 2.6. A.C. wave.

$$\text{Average induced back e.m.f.} = -L\frac{di}{dt}$$

$$= -L \times 4f I_m \text{ volts}$$

∴   Equal and opposite applied average e.m.f.

$$= 4f L I_m \text{ volts} \qquad (1)$$

But

Average value of sine wave = 0·637 × maximum value

$$= \frac{2}{\pi} V_m \qquad (2)$$

Since expressions (1) and (2) are equal

$$\frac{2}{\pi} V_m = 4f L I_m$$

∴

$$\frac{V_m}{I_m} = 2\pi f L$$

Expressing in r.m.s. values, $\dfrac{V}{I} = 2\pi f L$

Also $\dfrac{V}{I} = X_L$,   ∴ $\underline{X_L = 2\pi f L \text{ ohms}}$

2.7. Eddy Currents

E.m.f.s are not only induced in copper wires but in fact in any conducting material, such as the cores of machines or transformers, which undergo a change in flux-linkages. When induced in a mass of material e.m.f.s cause circulating currents to be produced, known as eddy currents.

Heat is set up by these currents passing through low resistance paths. The energy for the waste heat which is drawn from the mains has to be paid for and hence represents a loss. The temperature rise is also liable to damage the insulation of the windings. Use of laminations, lightly insulated from each other, rather than a solid bulk of conducting material, is one of the major means adopted to lessen the conducting path and hence the magnitude of the eddy currents. The loss and heat can also be further diminished if the stampings or laminations possess a high resistivity. Alloying iron with silicon increases resistivity and has a further beneficial effect in reducing the cyclic hysteresis loss.

2.8. Non-inductive Circuits

Eddy currents can also be induced in metal conduits which contain a live conductor only. For this reason, I.E.E. Wiring Regulation 521–8 states 'single-core cables armoured with steel wire or tape shall not be used for a.c. Conductors of a.c. circuits installed in ferrous enclosures shall be arranged so that the conductors of all phases and the neutral conductor (if any) are contained in the same enclosure. Where such conductors enter ferrous enclosures they shall be arranged so that the conductors are not separated by a ferrous material or provisions shall be made to prevent circulating eddy currents.'

For an indication of advances made in electrical installation safety techniques, it is instructive to compare with the equivalent 14th Edition B.100, 'Cables of a.c. circuits installed in steel conduits shall always be so bunched that the cables of all phases and the neutral conductor (if any) are contained in the same conduit'. This bunching neutralises the magnetic field so that there is no external flux to 'cut' the conduit in order to induce eddy currents.

Trouble has also been experienced by the eddy currents set up in the larger sizes of mineral-insulated cables, which can only be obtained in single-core types. Mystery heating has been traced to these single cables where placed in separately spaced runs.

In electrical measuring instruments, stray magnetic fields that induce e.m.f.s in the movement or instrument will affect the accuracy of the measuring device. Non-inductive coils are made by doubling back the winding (Fig. 2.7), so as to cancel out the magnetic field in each section of the wire.

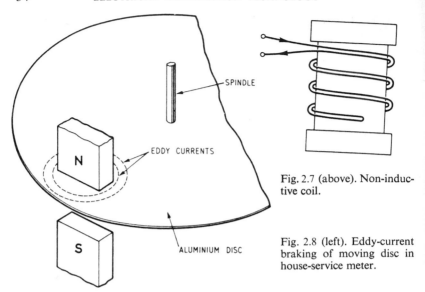

Fig. 2.7 (above). Non-inductive coil.

Fig. 2.8 (left). Eddy-current braking of moving disc in house-service meter.

Eddy currents are not entirely a nuisance. They can be turned to good account for braking purposes. Figure 2.8 illustrates part of a house-service meter, in which a flat metal disc rotates as an elementary motor, the spindle of the disc being attached to a clockwork train of gears for recording on dials the number of units consumed. When the supply being metered is turned off, the disc by itself would, due to inertia, rotate for some time before coming to rest. To brake the disc, it is placed between the poles of a permanent magnet. Any radius of the disc may be considered as a conductor as it passes the magnet. These 'linear conductors' cut the flux and produce eddy currents. By Lenz's Law they 'oppose the motion producing them' and this quickly brings the disc to rest.

Moving-coil instruments are rendered 'dead beat' on the same principle. The needle reaches its position rapidly with the minimum of oscillations due to the eddy currents induced in the aluminium former of the moving coil.

2.9. Mutual Inductance

As we have seen, a change in current through a conductor not only induces a back e.m.f. of self-induction in the conductor but in any other conductor linked by the changing flux. This effect is known as *mutual inductance* and is important in many ways. The transformer depends upon this action. Ignition coils, such as are used for producing the spark in motor-car engine cylinders, are also based on this principle.

Two coils have a mutual inductance of 1 H if a rate of change of current of 1 A/s in one coil induces an e.m.f. of 1 V in the other.

The symbol for mutual inductance is M.

Example 2.7

(i) *Calculate the mutual inductance between two adjacent coils if an average e.m.f. of 400 V is induced in one coil when a current of 20 A in the other is reversed in 0·1 second.*

(ii) *Two coils A and B have a mutual inductance of 10 mH. If the current in coil A is changing at a rate of 500 A/s, determine the e.m.f. induced in coil B.*

(i) Induced e.m.f. = $M \times$ rate of change of current $\left(\dfrac{di}{dt}\right)$

∴ $$M = \frac{\text{induced e.m.f.}}{\dfrac{di}{dt}}$$

Since a reversal of 20 A is equivalent to a current change of 40 A,

$$M = \frac{400 \text{ V}}{40/0\cdot1} = \frac{400 \text{ V}}{400 \text{ A/s}} = \underline{1 \text{ H}}$$

(ii) E.m.f. induced in coil B

$$= M\frac{di}{dt} = 0\cdot01 \times 500 = \underline{5 \text{ V}}$$

The coil where the supply current is changing is called the primary and the other the secondary. The e.m.f. induced in the latter opposes the change in the primary current similarly to self-inductance.

EXERCISES

1. State in which units *magnetic flux density* and *magnetising force* may be measured. Sketch a graph of magnetic flux density against magnetising force for (i) air, (ii) for iron.

Explain what is meant by the inductance of a coil and state with reasons, the effect on the inductance of an air-cored coil of (a) unwinding some of the turns from the coil (b) fitting the coil with an iron core.   [T]

2. Describe the way in which a small compass needle may be used to obtain a map of the magnetic field surrounding an electric current flowing round a coil of wire. Sketch the type of map you would expect to obtain, clearly

marking the direction of the current flow and the direction of the magnetic field.

Two similar coils A and B are wound on a straight iron bar. A d.c. supply is connected to coil A and a centre-zero galvanometer is connected to coil B. Describe the effect on the galvanometer when the current in coil A is (a) steadily increasing, (b) constant (c) suddenly reversed.

How would you obtain similar effects on the galvanometer with the aid of a bar magnet?   [T]

3. (a) Show clearly with the aid of a diagram the polarity of the e.m.f. produced when a conductor cuts a magnetic field at right angles. An aeroplane measures 25 m from wing tip to wing tip and is travelling horizontally at 330 miles/h. If the vertical component of the earth's magnetic field is 0·18 T, find the e.m.f. produced across the wing tips. (Hint—Induced e.m.f. = Blv volts, where v is the velocity in m/s.)

(b) Two long parallel conductors spaced 10 mm apart in air each carry a current of 100 A. Determine the force set up between the conductors per metre length.   [W. J.]

4. A coil with a number of tapping points is connected to a centre-zero galvanometer. A permanent magnet of bar form may be moved in and out of the coil.

Describe and give reasons for the deflections of the galvanometer pointer when

(a) the magnet is passed through the coil (i) slowly, and (ii) rapidly;

(b) the magnet is passed through the coil in the reverse direction (i) slowly, and (ii) rapidly;

(c) the magnet is held steady (i) within, and (ii) immediately outside the coil.

State the effect on the foregoing of (i) increasing the number of coil turns connected to the galvanometer, and (ii) using a more powerful magnet.

Note: Your answer should be illustrated by simple sketches, clearly showing the polarity of the magnet.

5. (a) Explain what is meant by the term self-inductance.

(b) List the factors on which the self-inductance of a coil may depend.

(c) The inductance of a circuit may be either an advantage or a disadvantage. Give reasons for your answers.

(d) If the current through a coil, having an inductance of 0·2 H changes uniformly from 8 A to 3 A in 0·01 second, calculate the value of the induced e.m.f.   [T]

6. Two coils are wound axially on a wooden former. The mutual inductance between them is 0·04 H. The current in one coil is allowed to increase from zero at a uniform rate of 50 A per second for a period of 5 milliseconds, remain constant for 10 milliseconds and then decrease at a uniform rate of 20 A per second to zero. Draw the waveform of this current and add (to scale) the waveform of the induced voltage in the second coil.

What effect would replacing the wooden former by an iron former have on the induced e.m.f.?

7. State Faraday's and Lenz's Laws.

A conductor 0·7 m long is moved at a uniform speed of 12 metres/second at right angles to its length and to a uniform magnetic field. If the e.m.f. generated in the conductor is 1·5 volt and the conductor forms part of a closed circuit having a resistance of 0·2 ohm, calculate: (i) the density of the magnetic field; (ii) the force acting on the conductor in newtons; (iii) the work done in joules when the conductor has moved 0·4 m.   [U]

8. Describe how eddy currents are set up. Give their disadvantages and state whether such currents can be put to good use.

Chapter 3

STATIC ELECTRICITY

Static electricity refers to electricity 'at rest' on an *insulator* in contrast to current electricity where the electrons are in continuous movement through a *conductor*.

A common effect of static electricity occurs when a portion of insulation being stripped from a p.v.c.-insulated cable flies off and clings to a wall. If a fountain pen is rubbed, it is found to pick up small pieces of light materials such as paper, etc. During dry cold winters, office workers walking on nylon carpets have been known to feel a prickle and even see sparks coming from their fingers when approaching hot-water radiators. These are all simple examples of electric charges that have been built up. An understanding of them helps to explain electrical applications of great practical importance.

3.1. Charges

The atoms which comprise all matter, i.e. everything which occupies space, are ultimately electrical in nature. The simple hydrogen atom indicates the general pattern. The inner core or nucleus contains the positively-charged *proton*; orbiting round the nucleus is the single negatively-charged electron.

Unimaginably small as is the atom—smaller than 10^{-6} millimetres in diameter—its structure contains the *neutron*, which possesses no charge, and many other sub-atomic particles. Fortunately, most of our electrical theory can be explained in terms of the proton and electron parts of the atom alone.

Normal materials, such as the book you are reading, are electrically neutral, i.e., they show no external electricity because the proton and electron charges exactly balance or cancel each other out. An atom is *positively charged when electrons are lost and negatively charged when an atom possesses surplus electrons*. Sometimes the process is known as *ionisation*. Atoms which are deficient in electrons are called positive ions and those which have gained electrons, negative ions.

Charges obey the fundamental law that like charges repel and unlike charges attract each other. This may be demonstrated by

38

suspended lightweight insulated balls such as pith balls when appropriately charged (Fig. 3.1). Ordinary balloons, as used for decorative purposes, when charged by rubbing can also be made to behave in this manner.

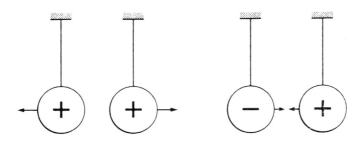

Fig. 3.1. Unlike charges attract, like charges repel.

3.2. Frictional Electricity

By rubbing a ballpoint pen (not the writing end) against the coat sleeve and placing it near small pieces of dry paper, the pieces will be attracted to the pen. Why is this? Electrons are rubbed off from the sleeve to the pen. When brought near to a piece of paper, the pen repels electrons from that part of the paper nearest to the pen, leaving a positive charge so that attraction takes place (Fig. 3.2). Quick slicing of p.v.c. cable may charge portions of the insulation

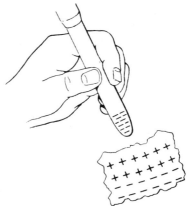

Fig. 3.2. A ballpoint pen produces an opposite charge in an adjacent small piece of paper, hence attraction occurs.

which come away so that they will often cling to a wall until the electron movement corrects the unbalance.

This redistribution of electrons is of great significance to electrical and technician engineers; as we shall see, there are also many important applications to electrical installation work. Dr. P. S. H. Henry, of the Cotton, Silk and Man-made Fibres Research Association, in a lecture to the A.S.E.E. pointed out that as more materials with higher electrical insulation were manufactured and used, particularly in processes involving continuous belts, the number of fires attributable to sparks from static electricity increased, earthing being the general solution.

Contrary to general belief, electrostatic electricity can arise without rubbing. When two surfaces are pressed lightly together, the electrons may intermingle and may even cross from one body to another to create an unbalance. Friction however increases both the area and degree of contact, so that this action is enormously amplified. Thus walking, particularly shuffling, along a nylon carpet can build up a charge on the person by creating such an unbalance. If the charges are strong enough, electrons may actually jump from the negatively-charged body to the oppositely-charged one. Under such circumstances sparks may be produced when approaching hot-water pipes or radiators.

3.3. Insulators and Conductors

Investigation shows that static charges *remain on the surface of insulators*. In fact this may be taken as the test of an insulator. If an uninsulated metal rod is rubbed no charge will be found on it. The charge has travelled or leaked away because of a movement of electrons through the metal as quickly as the charge has been produced. Insulators are thus materials which do not permit the flow of charges.

3.4. Electrostatic Induction

By electrostatic induction, the process of electrification can take place without even contact or rubbing. In the simplest form, the principle can be shown by a small metal ball placed on an insulated stand. A negative charge may be produced on an ebonite rod by rubbing and if the rod is brought near to the metal sphere redistribution of the charges on the ball will take place (Fig. 3.3(a)). This action is due to the fundamental electrostatic law as previously stated. The negatively-charged rod has a surplus of electrons; thus when it is brought near to the ball the electrons of the latter are repelled leaving

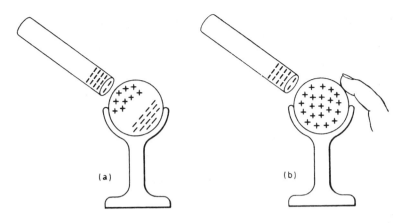

Fig. 3.3. Charging by induction. (*a*) Negatively-charged ebonite rod being used to redistribute charges on a metal ball. (*b*) Providing an escape path to earth for the repelled electrons.

the nearer side of the ball with a deficiency of electrons, i.e. positively-charged.

If the charged rod is removed, the sphere will revert to its neutral state. Connection to earth is normally used to disperse a charge, but in this case earthing may be employed to maintain the charge. If the ball is touched as shown in Fig. 3.3(*b*) while the negatively-charged rod is in the proximity of the ball, the repelled electrons have an escape path to earth. Removal of the finger then cuts off the path and the sphere remains positively-charged, since there is now a deficiency of electrons, even when the rod is taken away.

By bringing a positive charge near a neutral insulated metal body, the body can be made to possess a negative charge by the same process. In this case, the electrons will flow up from earth to the charged conductor.

The earth is presumed to be electrically neutral under all conditions, which is another way of saying that it possesses an almost unlimited reservoir of electrons.

3.5. Electrostatic Fields

Electrostatic fields can be plotted, but plotting is much more difficult to carry out than in the magnetic case. Two simple examples of such fields are shown in Fig. 3.4. *The direction of the electric lines of force or flux is the direction taken by a positive charge*, i.e. from

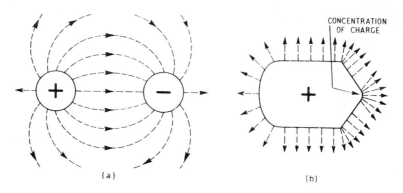

Fig. 3.4. (*a*) The direction of electrostatic lines are from positive to negative. (*b*) Charges tend to accumulate or concentrate at sharp points.

positive to negative (Fig. 3.4(*a*)). At any sharp points on an insulator there will be an accumulation or concentration of the charge (Fig. 3.4(*b*)). Note also, in contrast to magnetic lines, electrostatic lines start and end *on the surface*. If a positive and negative charge are connected by a wire, a displacement or movement takes place. The charges will even out until the parts concerned are both unelectrified (Fig. 3.5). The reader will notice that this is exactly what occurs when a capacitor is discharged by shorting out the terminals, except that the discharge will be seen as sparks if there are heavy charges on the plates. Light sparks may also appear, accompanied by tingling of the finger-tips when approaching radiators under the conditions already cited.

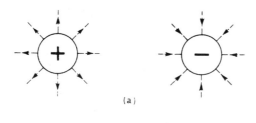

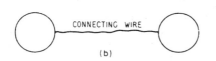

Fig. 3.5. Neutralising charges. (*a*) Separate positive and negative charge (note symmetry of field). (*b*) Electrical unbalance is evened out.

3.6. Potential and Potential Difference

Charged substances acquire potential energy and are said to possess a potential, or level of electricity, with respect to earth, earth being taken as being at zero potential. A positive charge is taken conventionally as being at a higher potential than a negative charge. Referring to Fig. 3.4(a), a potential difference (p.d.) exists between these two charges. Also, again conventionally, electricity flows from positive to negative when connected by the wire. By flow we mean a movement of electrons which in reality is from negative to positive. Connexion by the wire allows the surplus electrons from the negative charge to make up for the deficiency of electrons at the positive until each rapidly ceases to possess any charge.

3.7. Lightning Conductors

Lightning is as an example of the accumulation of gigantic static charges which may be set up on clouds by the movement ('rubbing') of air currents. Discharges to earth accompanied by massive arcing constitute lightning or lightning strokes. It is never advisable to stand under a tree during lightning since the tree and 'local' ground form part of the same conducting path. Vast energies are unleashed, destroying tall buildings when struck.

The purpose of the lightning conductor with its pointed spike is to provide a path of low resistance to the lightning current—which may be some 50 000 A—so that the discharge to earth will occur through the copper lightning conductor rather than through the building.

The actual discharge which reaches its peak value in several microseconds may consist of a series of strokes. With thunderstorms on 75 days per annum in Great Britain, lightning constitutes a potential major hazard to life and property, tall buildings being particularly vulnerable.

Lightning conductors essentially consist of three sections: (1) *Upper terminal*, which contrary to common belief need not be pointed and in fact may be fitted horizontally round the upper perimeter of a building. (2) *Copper tape* with the alternatives of galvanised steel or stranded conductors. It should be run as near vertical as possible and not in a lift shaft. Bonding should be made to metal projections such as gutters and there should be the minimum of joints. (3) *Earth electrode*, which may consist of metal rods driven deeply into the soil, or shorter rods or strips connected in parallel. The complete arrangement must be mechanically secure and have a maximum resistance of 10 Ω.

Ground areas afforded protection by the lightning conductor scheme are known as 'zones of protection'. Two simple cases are indicated in Figs. 3.6 and 3.7.

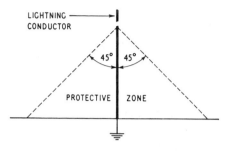

Fig. 3.6. Protection given by vertical lightning conductor.

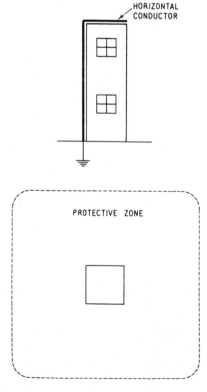

Fig. 3.7. Protection where vertical conductor linked to an horizontal section.

For fuller information the reader is referred to the British Standard Code of Practice CP 326 : 1965, *The Protection of Structures Against Lightning.*

3.8. Static Charges in Industry

Conveyors and motor-driven continuous belting often produce accumulation of static charges by friction. Under favourable conditions for the building up of charges, problems can arise due to the charged particles clinging together when oppositely charged, or flying apart when similarly charged. The particles may stick to machine parts and penetrate holes, a common feature being the attraction of dust particles.

Due to the possible creation of large charges, dangerous shocks may be felt by the operators of textile machines. Man-made fibres are particularly prone to this kind of charging. Explosive mixtures of fine dust can also be detonated by the accumulation of electrostatic charges.

Many chemical liquids in common use, such as petroleum spirit, have surprisingly high insulating qualities. When they are handled, or flowing out of pipes or set in violent agitation, very large static charges may be generated. Containers and pipes should be bonded and permanently earthed at as many points as possible to guard against a potential rise caused by the electrostatic process.

Earthing around the built-up charges normally serves as a simple solution to static problems; this allows the charges to leak away harmlessly to earth on the principle of the lightning conductor. A method sometimes employed is to make the materials more conductive by introducing some highly conductive substance into or on the surface of the material. Anti-static sprays can also be used, a film being formed which neutralises charges and acts to prevent further static from forming.

3.9. Hospital Static Charges

It is possible to build up quite a substantial charge by the movements of trolleys with rubber wheels on non-conducting floors. If these charges result in sparking they can be particularly dangerous in operating theatres, where anaesthetics such as ether or cyclopropane become explosive mixtures if combined with the oxygen of the air.

A cure is suggested by the use of metal chains but it is better to render the rubber wheels conductive by the introduction of carbon black. The floor itself can be made conductive by the addition of carbon black to the cement base over which a terrazo surface is laid, wire netting being incorporated in the floors.

3.10. The Capacitor

Almost any arrangement between two conducting bodies that are insulated from each other can come under this heading. It is however normal to restrict the term to two or more adjacent metal plates as symbolised by the two close vertical lines in Fig. 3.8. A capacitor has

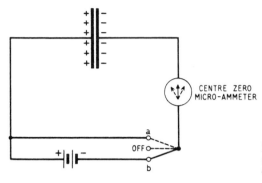

Fig. 3.8. Circuit illustrating capacitor charge and discharge.

the property of storing charges. As the 2-way switch makes contact with *b* from the off position, a momentary current passes, which is indicated by the needle of the microammeter moving in one direction. A p.d. is found to exist between the capacitor plates, with polarities as shown. If the switch is moved to the off position the plates do not lose their charge, showing that these charges have been stored. Discharge occurs when the switch is moved over to *a*, as shown by movement of the microammeter needle in the opposite direction. The discharge takes place because the p.d. causes a current to pass from one plate to the other. Note that no current goes through, i.e. between, the plates. The current movement is more in the nature of an oscillation (to and fro movement) than a continuous flow.

3.11 Charge and Discharge Curves

CHARGE AND DISCHARGE CURVES. By plotting charge and discharge currents against time (Fig. 3.9) we can appreciate more clearly what occurs. With a resistance in circuit and the battery connected to the capacitor, the initial current as shown by the microammeter immediately rises to its maximum value E/R, where E is the battery e.m.f. and R the circuit resistance.

As the capacitor becomes charged, a rising p.d. appears across the plates. The direction of this p.d. is such as to oppose the battery e.m.f.

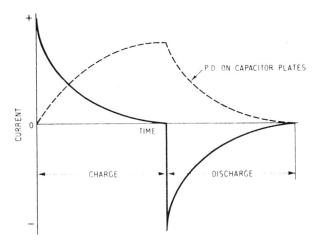

Fig. 3.9. Capacitor charge and discharge curves.

Because of this rising opposition the current falls and becomes zero when the capacitor p.d. is equal to the battery voltage. The capacitor is then said to be charged. The relationship between the current and voltage can now be seen as a 'current lead'. When the current is at its maximum value and positive, the p.d. is zero. As the current reduces to zero, the p.d. reaches its maximum value.

On discharge, the current immediately rises to the same magnitude as in charge, but it flows in the opposite direction. It falls away to zero as the p.d. drops to zero.

As in the case of an inductor in series with a resistor (Section 2.2) the value of the time constant is necessary for the graphical representation of the charge and discharge curves. Here $T = CR$ s, where C is the capacitance in farads and R is the resistence in ohms; it is the time taken to reach full voltage across the capacitor if continued at the same commencing rate.

Example 3.1

A 15 μF capacitor is charged to a potential difference of 200 V and then discharged through a 50 kΩ resistor.

(a) Calculate (i) the value of the time constant, and (ii) the initial value of the discharge current with respect to time from switching on.

(b) Construct a graph showing the decay of the discharge current with respect to time from switching on.

(*c*) *From the graph find the value of the current after a time of* 1·5 *s.*

[C]

(*a*)(i)   Time constant $= CR$

$$= \frac{15 \times 50 \times 10^3}{10^6} \qquad = \underline{0{\cdot}75 \text{ s}}$$

(ii)   Initial value of discharge current

$$= \frac{V}{R}$$

$$= \frac{200 \times 10^3}{50 \times 10^3} \qquad = \underline{4 \text{ mA}}$$

(*b*) The graphical representation of decaying current is shown in Fig. 3.10.

(*c*) From the graph, the value of the current after 1·5 s = $\underline{0{\cdot}4 \text{ mA}}$

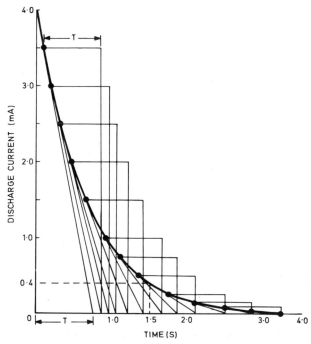

Fig. 3.10. Curve of capacitor discharge current.

The property of storing a charge is known as *capacitance*. We should bear in mind that capacitance is not entirely confined to the parallel-plate arrangement. Cables, especially long lines, when in close proximity possess this property, which may have to be taken into account.

3.12. Units

The charge of a single electron, although exceedingly small, forms the fundamental unit of quantity of electricity. The COULOMB (abbreviation C) is the practical unit which can be taken as being equal to a charge of 6.2×10^{18} electrons. As electric *current* is the *rate of flow* of electricity, therefore

$$1 \text{ ampere} = 1 \text{ coulomb/second}$$

$$= 6.2 \times 10^{18} \text{ electrons passing a given point in 1 second (Fig. 3.11)}$$

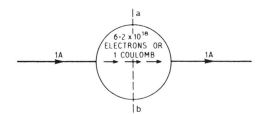

Fig. 3.11. Relation between coulomb (quantity of electricity) and ampere (current flow). There is a current of 1 A when 6.2×10^{18} electrons, or 1 coulomb, passes section *ab* in one second.

The quantity in coulombs which a capacitor will store is proportional to the charging voltage. Capacitance as a unit is defined as the quantity stored in coulombs when 1 volt is applied across the plates.

A capacitor possesses a capacitance of 1 farad (F) if an applied potential difference of 1 volt across the plates is capable of storing 1 coulomb.

Hence
$$C = \frac{Q}{V}$$

where Q is in coulombs and C in farads. It may be easier to remember in the form

$$Q = VC$$

For practical purposes the farad is much too large, therefore the microfarad (μF), which is 1 millionth part of a farad, is used. For smaller capacitors $\mu\mu F$ has been replaced by picofarad (pF) which is equal to 10^{-12} F.

Example 3.2

(i) *An 8 μF capacitor is connected to a p.d. of 250 V. Calculate the charge on plates.*

(ii) *A certain capacitor requires a charge of 100 microcoulombs to produce a p.d. of 75 V between the capacitor plates. What is the value of the capacitance of the capacitor?*

(i) $Q = VC = \dfrac{250 \times 8}{10^6}$ $= \underline{2000 \ \mu C}$ (microcoulombs)

(ii) By transposition,

$$C = \frac{Q}{V} = \frac{100 \times 10^{-6}}{75} \qquad = \underline{1 \cdot 33 \ \mu F}$$

3.13. Capacitors in Series

With capacitors connected in this way and referring to Fig, 3.12, the polarities on the plates will be as shown. Since a similar electron movement occurs at each plate, the same quantity of electricity is stored by each capacitor when in series.

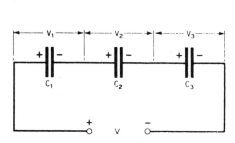

Fig. 3.12. Capacitors in series.

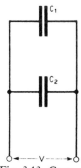

Fig. 3.13. Capacitors in parallel.

Clearly the supply voltage is equal to the sum of the individual voltages.

i.e. $\qquad\qquad\qquad V = V_1 + V_2 + V_3$

∴ $\qquad\qquad\qquad \dfrac{Q}{C} = \dfrac{Q}{C_1} + \dfrac{Q}{C_2} + \dfrac{Q}{C_3}$

Dividing through by Q

$$\frac{1}{C} = \frac{1}{C_1} + \frac{1}{C_2} + \frac{1}{C_3}$$

Thus the reciprocal of the equivalent capacitance is equal to the sum of the reciprocal of the individual capacitances when in series.

Example 3.3

A circuit consists of 1, 2, 3 and 4 μF capacitors in series. What is the value of the eqivalent capacitance?

$$\frac{1}{C} = \frac{1}{C_1} + \frac{1}{C_2} + \frac{1}{C_3} + \frac{1}{C_4}$$

$$= \frac{1}{1} + \frac{1}{2} + \frac{1}{3} + \frac{1}{4} = \frac{12 + 6 + 4 + 3}{12} \qquad = \frac{25}{12}$$

$$\therefore \qquad C = \frac{12}{25}\mu F \qquad \qquad = 0.48 \ \mu F$$

Note that, in a similar way to resistors in parallel, the equivalent capacitance of capacitors in series is less than any single capacitance. A simplified form for two capacitors in series may be taken as:

$$\frac{1}{C} = \frac{1}{C_1} + \frac{1}{C_2} \quad \text{i.e.} \quad C = \frac{C_1 C_2}{C_1 + C_2}$$

Thus for 5 μF and 10 μF in series, the equivalent capacitance is given by

$$\frac{5 \times 10}{5 + 10} \quad \text{i.e.} \quad \underline{3.33 \ \mu F}$$

3.14. Capacitors in Parallel

The same voltage V exists across each capacitor (Fig. 3.13).

$$\text{Charge on } C_1, Q_1 = VC_1$$

$$\text{Charge on } C_2, Q_2 = VC_2$$

But total charge $Q = VC$ where C is the equivalent capacitance

Also

$$Q = Q_1 + Q_2$$

$$VC = VC_1 + VC_2$$

Dividing through by V,

$$C = C_1 + C_2.$$

The total capacitance of two (or more) capacitors in parallel = the sum of the individual capacitances.

3.15. Capacitors in Series-parallel

Example 3.4

In the accompanying circuit (Fig. 3.14) the total charge is 750 μC. Determine the value of V_1, V and C_2.

$$V_1 = \frac{Q}{C_1} = \frac{750}{15} \qquad\qquad = \underline{50\ V}$$

$$V = V_1 + V_2 = 50\ V + 20\ V \qquad = \underline{70\ V}$$

$$\text{Charge on } C_3 = 20\ V \times 8\ \mu F \qquad = 160\ \mu C$$

$$\text{Charge on } C_2 = 750\ \mu C - 160\ \mu C = 590\ \mu C$$

$$\text{Capacitance of } C_2 = \frac{590\ \mu C}{20\ V} \qquad\qquad = \underline{29 \cdot 5\ \mu F}$$

As a check, the student could work backwards from the value as calculated and ascertain whether the total charge is 750 μC.

3.16. Electric Flux, Flux Density

Consider a p.d. of V volts applied to the two plates of a capacitor (Fig. 3.15). With polarities as shown, the left-hand plate will experience a positive charge of Q coulombs and the right-hand plate a charge of the same magnitude but of opposite polarity. There is electric flux or electric lines of force between the plates, the number

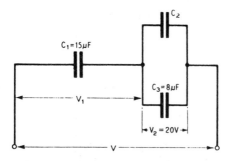

Fig. 3.14. Capacitors in series-parallel. (For calculations, see Example 3.4).

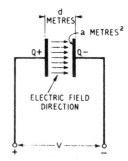

Fig. 3.15. Two-plate capacitor.

of electric lines being taken as numerically equal to the charge in coulombs on each plate.

$$\text{Electric flux density } (D) = \frac{\text{No. of lines of electric flux}}{\text{Area of one plate in m}^2}$$

$$= \frac{Q}{a} \text{ coulombs/m}^2$$

3.17. Electric Force

Referring to Fig. 3.15, electric force or electric field strength (E) equals V/d volts per metre. This quantity is sometimes called *potential gradient* and is of great importance in considering the breakdown strength of a capacitor. Too great a voltage will produce a spark across the gap and the insulator between the plates will be punctured. Capacitors are marked with the maximum working voltage and it is important to ensure that this value is not exceeded.

3.18. Permittivity

By now the reader may have understood that a number of electrostatic relationships and concepts bear a striking similarity to those used for magnetic calculations. For example,

$$\mu \text{ (permeability)} = \frac{B \text{ (flux density)}}{H \text{ (magnetising force)}}$$

Here,

$$\varepsilon \text{ (permittivity)} = \frac{D \text{ (electric flux density)}}{E \text{ (electric force)}}$$

Also $\varepsilon = \varepsilon_0 \varepsilon_r$ where $\varepsilon_0 = 8 \cdot 85 \times 10^{-12}$ F/m

and $\varepsilon_0 = $ absolute permittivity

$\varepsilon_r = $ relative permittivity

(Note: ε is the Greek letter, epsilon.)

The value of the relative permittivity depends upon the material used for the dielectric between capacitor plates. For this reason, different charges will be stored for the same applied p.d. and dimensions by changes in the nature of the material.

3.19. Capacitor Dimensions

Now
$$\frac{D}{E} = \frac{Q/a}{V/d} = \frac{Q}{a} \times \frac{d}{V}$$

Rearranging
$$\frac{D}{E} = \frac{Q}{V} \times \frac{d}{a} \quad \text{but} \quad \frac{Q}{V} = C$$

$$= C \times \frac{d}{a} \quad \text{also} \quad \frac{D}{E} = \varepsilon$$

$$\therefore \qquad \varepsilon_0 \varepsilon_r = C \times \frac{d}{a}$$

By transposition,

$$C = \varepsilon_0 \varepsilon_r \times \frac{a}{d}$$

$$= \frac{8 \cdot 85 \times 10^{-12} \times \varepsilon_r \times a}{d} \text{ farads}$$

From this expression we can draw the conclusion that the capacitance is proportional to the area of the plates (a) and inversely proportional to the distance between plates (d).

Example 3.5

Each plate of a 2-plate capacitor has an area of 200 cm². The whole of the space between the plates which are spaced 2 mm apart is filled with an insulating material having a relative permittivity of 4. If a potential difference of 500 V is applied across the plates, determine
(a) the capacitance of the arrangement,
(b) the charge stored,
(c) the electric flux density and the electric force.   [E]

(a) Capacitance,

$$C = \frac{\varepsilon_0 \varepsilon_r a}{d}$$

$$= \frac{8 \cdot 85 \times 10^{-12} \times 4 \times 200 \times 10^{-4} \times 10^6}{2 \times 10^{-3}}$$

$$= \underline{0 \cdot 000354 \ \mu\text{F} \ (354 \ \text{pF})}$$

(b) Charge,

$$Q = VC$$

$$= 500 \times 0.000354$$

$$= 0.177 \text{ microcoulombs } (\mu C)$$

(N.B. Avoid confusion between the use of C as the symbol for capacitance and C the abbreviation of coulombs.)

(c) Flux density,

$$D = \frac{Q}{a} = \frac{0.177}{200 \times 10^{-4}} = 8.85 \ \mu C/m^2$$

Electric force,

$$E = \frac{V}{d} = \frac{500}{2 \times 10^{-3}} = 250\ 000 \text{ V/m}$$

3.20. Energy Stored in a Capacitor

With a potential difference across the plates current will flow, therefore energy (VIt) will be drawn from the source and remains on the plates until discharge takes place.

$$Q = VC \quad \text{coulombs} \quad \text{where } V \text{ is in volts and } C \text{ is in coulombs}$$

Also

$$Q = It \qquad \text{where } I = \text{average charging current in amperes and } t = \text{time in seconds}$$

∴ Charging current $\dfrac{Q}{t} = \dfrac{VC}{t}$

Referring to Fig. 3.16,

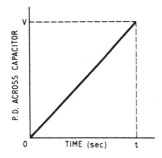

Fig. 3.16. Graph showing time taken for a p.d. across the capacitor plates to reach its final value.

Average p.d. across capacitor

$$= \frac{V-0}{2}$$

$$= \tfrac{1}{2}V \text{ volts}$$

∴   Average power during charge

$$= \text{average p.d.} \times \text{charging current}$$

$$= \frac{V}{2} \times \frac{VC}{t}$$

$$= \frac{V^2 C}{2t}$$

∴   Energy stored in a charged capacitor

$$= \text{power} \times \text{time}$$

$$= \frac{V^2 C}{2t} \times t \text{ joules}$$

$$= \tfrac{1}{2}V^2 C \text{ J}$$

Example 3.6

Two parallel-plate capacitors, C_1 and C_2 are connected in series across a d.c. supply of V volts. Show that the p.d. across C_1 is given by
$V_1 = \left(\dfrac{C_2}{C_1 + C_2}\right) V.$ *C_1 has a plate area of 0·195 m^2, a plate separation of 0·5 mm and the relative permittivity of the dielectric is 2·5. C_2 has a capacitance 0·03 μF. Calculate the total p.d. across two capacitors that will give rise to an electric force of 115 V/mm in the dielectric of C_1.*

$$\frac{1}{\text{Equivalent capacitance}} = \frac{1}{C} = \frac{1}{C_1} + \frac{1}{C_2}$$

∴ $C = \dfrac{C_1 C_2}{C_1 + C_2}$ (on the lines of two resistors in parallel)

Also $Q = VC$

and $V_1 = \dfrac{Q}{C_1} = \dfrac{VC}{C_1}$

$$= \frac{V \times C_1 C_2}{C_1(C_1 + C_2)} \qquad\qquad = \left(\frac{C_2}{C_1 + C_2}\right) V$$

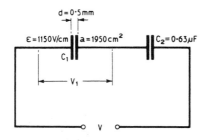

Fig. 3.17. Capacitors in series, Example 3.6.

Referring to Fig. 3.17,

Capacitance $\quad C = \dfrac{\varepsilon_0 \varepsilon_r a}{d}$

$$C_1 = \frac{8\cdot85 \times 10^{-12} \times 2\cdot5 \times 0\cdot195}{0\cdot5 \times 10^{-3}}$$

$\qquad\qquad = 8628 \times 10^{-12} \qquad\qquad\qquad = 0\cdot008628\ \mu F$

Electric force $\quad E = \dfrac{V}{d}$

$\qquad V_1 = Ed = 115 \times 0\cdot5 \times 10^3 \times 10^{-3} \;=\; 57\cdot5\ V$

Total capacitance

$$C = \frac{C_1 C_2}{C_1 + C_2} = \frac{0\cdot008\ 628 \times 0\cdot03}{0\cdot008\ 628 + 0\cdot03} \;=\; 0\cdot0067\ \mu F$$

$\quad Q = V_1 C_1 = 57\cdot5 \times 0\cdot008\ 628 \qquad\qquad = 0\cdot4962\ \mu C$

Total p.d. $\quad V = \dfrac{Q}{C} = \dfrac{0\cdot4962}{0\cdot0067} = 74\cdot03 \qquad\qquad = \underline{74\ V}$

3.21. Capacitors in A.C. Circuits—Capacitive Reactance

Where there is a continuous voltage variation, as produced by a.c. supplies, a continuous back e.m.f. is produced on the plates of the capacitor. This opposition to the change in current is known as Capacitive Reactance (X_C) and is measured in ohms.

$$X_C = \frac{10^6}{2\pi f C}\,\Omega \quad \text{where } C \text{ is in } \mu F$$

3.22. Derivation of $X_C = 1/2\pi f C$

The reasoning is somewhat similar to that employed for deriving inductive reactance (Chapter 2)

If frequency = f Hz, then time for 1 cycle = $1/f$ second.
∴   Time to reach maximum value ($\frac{1}{4}$ cycle) = $1/4f$ second.

Average rate of change of voltage

$$= \frac{\text{voltage increase}}{\text{time taken}}$$

Average rate of change of voltage during $\frac{1}{4}$ cycle

$$= \frac{V_m}{1/4f} \text{ where } V_m = \text{maximum value of voltage}$$

$$= 4f V_m \text{ volts/second}$$

Now,   Charge on capacitor plates in coulombs

$$= \text{volts across plates} \times \text{capacitance in farads}$$

i.e. $Q = VC$

∴   Current × time = C × volts

∴   Current = $C \times \dfrac{\text{volts}}{\text{time}}$

$$= C \times \text{rate of change of voltage}$$

∴   Average rate of change of current

$$= C \times \text{average rate of change of voltage}$$

∴ $\dfrac{2}{\pi} I_m = 4Cf V_m$

By transposition, $\dfrac{V_m}{I_m} = \dfrac{1}{2\pi f C}$

Expressing in r.m.s. values,

$$\frac{0 \cdot 707 V}{0 \cdot 707 I} = \frac{1}{2\pi f C}$$

$$\therefore \qquad \frac{V}{I} = \frac{1}{2\pi f C} \, \Omega$$

Also $\dfrac{V}{I} = X_C$ $\qquad \therefore \quad X_C = \dfrac{10^6}{2\pi f C} \, \Omega$ where C is in μF

EXERCISES

1. (a) Describe briefly the construction and operation of an electrostatic voltmeter. In what circumstances would this type of voltmeter be used?

(b) Three capacitors of values $8 \, \mu$F, $12 \, \mu$F and $16 \, \mu$F respectively are connected across a 240-V d.c. supply, (i) in series, and (ii) in parallel. Calculate in each case, the resultant capacitance, and also the potential difference across each capacitor.

(c) Why is it necessary to fit a discharge resistance to a capacitor? [C]

2. (a) Describe briefly how unwanted static electricity is produced in hospital operating theatres and two ways it arises in industry.

(b) Describe briefly how explosions may arise therefrom.

(c) Name (i) the most common method for ensuring safe dissipation of such static charges, and (ii) the type of switches that should be used where anaesthetic vapours may be present in hospitals. [C]

3. Derive an expression for the electrical energy stored in a capacitor of capacitance C farads charged to a p.d. of V volts. A capacitor having a capacitance of $3 \, \mu$F is charged to a p.d. of 200 V and then connected in parallel with an uncharged capacitor having a capacitance of $2 \, \mu$F. Calculate the p.d. across the parallel capacitors and the energy stored in each capacitor before and after being connected in parallel. Account for the difference in total energy before and after connexion.

4. (a) What is the unit of electric charge, and how is it related to the unit of electric current?

(b) A capacitor is charged at a constant rate of 10 mA for 30 minutes and then discharged at a constant rate in 1 second. What is the value of the current during discharge? [T]

5. A capacitor is composed of two flat parallel plates separated by a slab of insulation. Explain why a current will flow for a short time when this capacitor is connected across a d.c. supply.

In such a capacitor each of the plates is a rectangle of 100 mm × 50 mm, the thickness of the insulation is 5 mm and the total capacitance is $0 \cdot 0005 \, \mu$F. The capacitor is connected to a 100-V d.c. supply for some time. Calculate the dielectric flux density and the electric stress in the insulation, clearly stating the units in which each is measured.

What would be the capacitance of this arrangement if the thickness of the insulation were to be increased to 10 mm? [T]

6. What is meant by the *permittivity* of dielectric material?

A parallel-plate capacitor of C farads is charged by the transfer of Q coulombs from one plate to the other. Write down a formula for the potential difference V volts between the plates.

Capacitors of 1 μF and 2 μF are connected in parallel, and this combination is then connected in series with a capacitor of 6 μF.

Calculate the total capacitance of the circuit.   [T]

7. (*a*) Explain the term *relative permittivity*.

(*b*) A 0·001-μF capacitor is constructed from two metal plates each having an effective area of 500 cm^2 and spaced 1 mm apart. If a p.d. of 200 V is maintained between the plates calculate, giving clearly the units in each case:

 (*i*) the charge on the capacitor,

 (*ii*) the electric flux density between the plates,

 (*iii*) the electric force or potential gradient between the plates.

If the plate area is reduced by 20%, calculate the new distance required between the plates so that the capacitance may remain constant at 0·001 μF.

[T]

8. State a means of safeguarding against the effects of an accumulation of electric charges in the following: (*a*) chemical liquids through pipes, (*b*) conveyor belting. Why is the magnitude of these charges greater in dry than in damp weather?

9. (*a*) Explain with the aid of sketches, the meaning of the term 'zone of protection' when referring to a single vertical lightning conductor.

(*b*) Describe *one* method of providing lightning protection for a brick built chimney some 100 m high with diameter 2·5 m.

10. (*a*) A capacitor is connected in a d.c. circuit with a resistor across its terminals.

 (i) What is the purpose of the resistor?

 (ii) Sketch a graph showing the potential across the capacitor terminals against time.

(*b*) Two capacitors of 3 μF and 6 μF respectively are connected (i) in series, and (ii) in parallel. In each case a d.c. supply of 150 V is connected across the ends of the combination. For *each* circuit, find the resultant capacitance and the potential difference across *each* capacitor.

Chapter 4

D.C. CIRCUITS

In order to deal in a confident manner with more involved circuits, students should make a point of fully understanding simple series and parallel circuits. We commence with the 'rules of the game', but as in most parts of this work plenty of practice from earlier years' work will aid in achieving a permanent grasp of the subject.

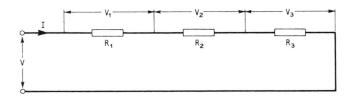

Fig. 4.1. The series circuit.

4.1. Resistors in Series

The supply voltage V is obviously equal to the sum of the potential difference across each resistor (Fig. 4.1):

Voltage law $\qquad V = V_1 + V_2 + V_3$

As the *same* current flows through each of the resistors,

$$V_1 = IR_1 \qquad V_2 = IR_2 \quad \text{and} \quad V_3 = IR_3$$

∴ $\qquad V = IR_1 + IR_2 + IR_3$

$$= I(R_1 + R_2 + R_3)$$

But $\qquad V = IR \quad$ where

$\qquad R = \text{total resistance}$

∴ $\qquad IR = I(R_1 + R_2 + R_3)$

thus $\qquad R = R_1 + R_2 + R_3$

Example 4.1

Four similar indicator filament lamps, each rated at 5 W, 50 V, are connected in series across a 200 V supply. What is the total current taken from the supply?

After a period of operation, one of the lamps fails and becomes open-circuited. Explain how a voltmeter may be used to find which lamp has failed, stating clearly what readings would be expected on the voltmeter.

The only replacement lamp available is one rated at 2·5 W 50 V. What would be the voltages across this lamp and each of the other lamps if this replacement were used in the circuit, and what would be the probable result?   [T]

$$\text{Power in watts} = \text{voltage} \times \text{current}$$

$$P = VI$$

Substituting $\dfrac{V}{R}$ for I

$$P = \frac{V^2}{R}$$

By transposition $R = \dfrac{V^2}{P}$

Thus, resistance of each of the 5 W, 50 V lamps

$$= \frac{50 \times 50}{5}$$ $= 500\ \Omega$

Total resistance of the 4 lamps $= 2000\ \Omega$

Current in circuit $= \dfrac{200\text{ V}}{2000\ \Omega}$ $= \underline{0\cdot1\text{ A}}$

Since an open-circuited lamp does not allow any current to pass in a series circuit

Reading by voltmeter across good lamps = zero

Reading across faulty lamp = practically full
voltage of 200 V

This is because the voltmeter would have a high resistance, being 1000 Ω/V for a good instrument, i.e. $200 \times 1000 = 200\,000\ \Omega$, which is now placed in series with the resistance of the three lamps totalling 1500 Ω.

Voltage is given by $I \times R$, and since in a series circuit the current is the same for all parts of the circuit, almost the whole of the 200 V supply will appear across the voltmeter.

$$\text{Resistance of 2·5 W, 50 V lamp} = \frac{V^2}{P} = \frac{50 \times 50}{2·5} = 1000 \ \Omega$$

$$\text{Total resistance is now} = 1500 + 1000 = 2500 \ \Omega$$

By proportion, the voltage across replacement lamp

$$= \frac{1000}{2500} \times 200 = 80 \text{ V}$$

The lamp would probably burn out after a short period.

(*Comment.* A common error is to add the current taken by each lamp when rated at 50 V. The current in the circuit can only be found by dividing the total resistance of the lamps into the supply voltage.)

4.2. Fall of Potential

Potential is often stated as referring to the *level of electricity*. When a battery delivers current, a potential difference exists between

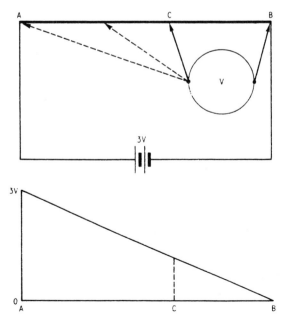

Fig. 4.2. Fall of potential along a wire. Point C at a higher potential than B but lower than A.

the terminals. Since the conventional direction of current gives a flow from positive to negative, positive terminal is said to be at a higher potential than negative (Fig. 4.2). When the movable contact C is moved towards A, the voltmeter will show an increasingly higher reading. If wire AB forms a uniform conductor the voltmeter reading increases strictly in proportion to its contact distance from B. From the graphical representation it can also be seen that at any intermediate point C is at a lower potential than A but at a higher potential than B. With B connected to earth then A would be at a potential *above* earth. These ideas of potential are crucial in understanding the 3-wire d.c. method of distribution.

Example 4.2

(a) *One end of an 8 Ω resistor is connected to the positive terminal of a battery X having an e.m.f. of 10 V. The other end is connected to a negative pole of a battery Y of e.m.f. 4 V, the positive pole of the latter being connected to the negative pole of battery X by means of a 2 Ω resistor. Assuming the batteries have negligible resistance, and the negative pole of X is earthed, determine*

    (i) *the current flow,*

    (ii) *the p.d. across each resistor.*

(b) *Represent the circuit conditions by means of a graph.*

(a) The circuit may be drawn as Fig. 4.3.

    (i)   Total e.m.f. $= 10 + 4 = 14$ V

$$\text{Total resistance} = 8 + 2 = 10 \, \Omega$$

$$\text{Current} = \tfrac{14}{10} \quad = \underline{1 \cdot 4 \, \text{A}}$$

    (ii)   Potential difference $= IR$

The potential difference across the 8 Ω resistor

$$= 1 \cdot 4 \times 8 = \underline{11 \cdot 2 \, \text{V}}$$

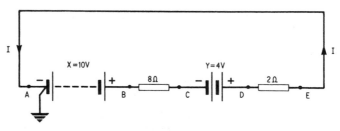

Fig. 4.3. Resistors and batteries in series.

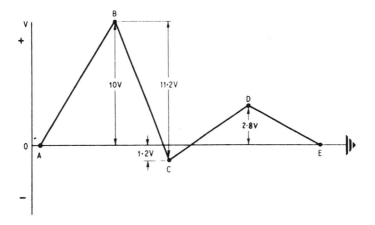

Fig. 4.4. Rise and fall of potential along a circuit.

The potential difference across the $2\,\Omega$ resistor

$$= 1\cdot4 \times 2 = \underline{2\cdot8\text{ V}}$$

(b) With the circuit as illustrated in Fig. 4.3, there is a rise in potential from A to B and a fall from B to C along the $8\,\Omega$ resistor. From C to D a potential rise occurs due to the e.m.f. in the 4 V battery. There is a final fall of potential along the $2\,\Omega$ resistor from D to E.

The graph in Fig. 4.4 shows that with both A and E at earth potential, B is 10 V *above* earth, C is $11\cdot2$ V—10 V, i.e. $1\cdot2$ V *below* earth potential, and D is $2\cdot8$ V *above* earth potential.

4.3. Parallel Circuits

In the basic parallel circuit (Fig. 4.5), the main current I is the sum of the individual branch currents.

Current law $\qquad\qquad I = I_1 + I_2 + I_3$

Since the voltage V is common and $I = \dfrac{V}{R}$

$$\frac{V}{R} = \frac{V}{R_1} + \frac{V}{R_2} + \frac{V}{R_3}\qquad \text{where } R \text{ is the equivalent resistance of the group}$$

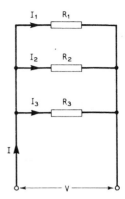

Fig. 4.5. Currents through resistors connected in parallel.

Dividing through by V,

$$\frac{1}{R} = \frac{1}{R_1} + \frac{1}{R_2} + \frac{1}{R_3}$$

Occasionally, this reciprocal form can be usefully replaced by

$$R = \frac{R_1 R_2 R_3}{R_1 R_2 + R_1 R_3 + R_2 R_3}$$

The expression above is obtained through using the product $R_1 R_2 R_3$ as an L.C.M. More often, for two resistors in parallel, the expression

$$R = \frac{R_1 R_2}{R_1 + R_2}$$

makes for easier working.

For series-parallel circuits, resistors in parallel should be seen *as a single group* in series with other resistors.

Example 4.3

Four resistors, AB, BC, AD and DC, are connected together to form a closed square ABCD. The known resistance values are: AD 12 Ω, AB 35 Ω, and DC 12 Ω.

A d.c. supply of 120 V is connected to A and C so that current enters the combination at A and leaves at C. A high-resistance voltmeter is connected between B and D and, whilst carrying negligible current, registers a voltage drop of 10 V from B to D.

(a) Calculate the value of the resistance BC, and the total current taken from the supply.

(b) Calculate also the value of BC, such that the potential difference between B and D is in the reverse direction i.e. from D to B.   [C]

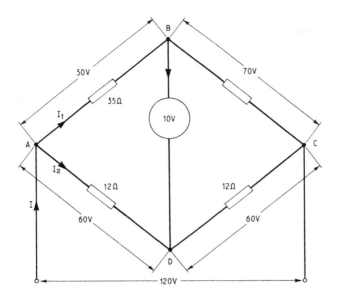

Fig. 4.6. Circuit conditions of balanced bridge.

(*a*) The circuit conditions are given in Fig. 4.6.

The supply current *I* enters the junction at A and splits up into the currents I_1 and I_2 as shown; also the network consists of two parallel branches each comprising two resistors in series, since negligible current passes through the voltmeter.

As the resistors AD and DC are equal and form one parallel branch, a p.d. of 60 V must exist between the ends of these resistors.

Also, because D is at a potential of 60 V with respect to A, point B must be at a potential of 50 V with respect to A in order to produce a voltage drop from B to D. Thus, the p.d. across BC = 70 V.

Now
$$I_1 = \frac{50 \text{ V}}{35 \text{ }\Omega} = 1\tfrac{3}{7} \text{ A}$$

$$I_2 = \frac{60 \text{ V}}{12 \text{ }\Omega} = 5 \text{ A}$$

The total current taken from the supply
$$I = I_1 + I_2 = 1\tfrac{3}{7} + 5 \qquad\qquad = \underline{6\tfrac{3}{7} \text{ A}}$$

Because the voltmeter is of very high resistance, I_1 also flows through BC and

$$\text{the resistance of BC} = \frac{70 \text{ V}}{1\frac{3}{7} \text{ A}} = 49 \, \Omega$$

(b) The voltage values across AB and BC are now reversed.

$$\therefore \quad \text{The new value for } I_1 = \frac{70 \text{ V}}{35 \, \Omega} = 2 \text{ A}$$

and

$$\text{the new value of resistor BC} = \frac{50 \text{ V}}{2 \text{ A}}$$

$$= 25 \, \Omega$$

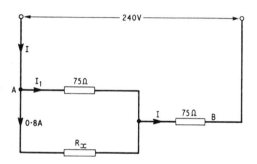

Fig. 4.7. Equivalent series-parallel circuit.

Example 4.4

A 150 Ω resistance coil AB is connected to terminals at 240 V d.c. Calculate the value of a further resistance coil which, when connected between the mid-point of AB and end A, will carry a current of 0·8 ampere. [C]

The circuit conditions may be shown by an equivalent diagram (Fig. 4.7). In the diagram

$$I = \text{main current}$$

$$R_x = \text{resistance to be determined}$$

Now,   Voltage across series resistor $= 75I$

$\therefore$ Voltage across parallel group $= 240 - 75I$

The main current is equal to the sum of the currents in the parallel group,

i.e. $I = I_1 + 0.8$

∴ $I = \dfrac{240 - 75I}{75} + 0.8$

∴ $75I = 240 - 75I + 0.8 \times 75$ when multiplied throughout by 75

∴ $150I = 240 + 60$

∴ $I = 2\,\text{A}$

∴   Voltage across parallel group

$$= 240 - 75 \times 2 = 90\,\text{V}$$

∴ $R_x = \dfrac{90\,\text{V}}{0.8\,\text{A}}$ $= \underline{112.5\,\Omega}$

4.4. Branch Current Law

To find the branch currents, I_1 and I_2, in a parallel circuit (Fig. 4.8), students usually start by calculating the voltage, V. This step can be avoided when the main current is known by employing the expressions:

$$I_1 = I \times \frac{R_2}{R_1 + R_2} \qquad I_2 = I \times \frac{R_1}{R_1 + R_2}$$

These may appear involved, but the example following the proof shows that they are extremely easy to use. There is also the further advantage that the equations help to drive home the basic theory that the distribution of currents in a parallel circuit is inversely proportional to the resistance in each branch.

PROOF. As previously stated the equivalent resistance of two resistors in parallel

$$R = \frac{R_1 R_2}{R_1 + R_2}$$

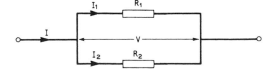

Fig. 4.8. Distribution of currents.

Referring to Fig. 4.8, $V = IR$

i.e. $= I \times \dfrac{R_1 R_2}{R_1 + R_2}$

But $I_1 = \dfrac{V}{R_1}$

Substituting for V $I_1 = \dfrac{I \times \dfrac{R_1 R_2}{R_1 + R_2}}{R_1}$

Dividing numerator and denominator of right-hand side by R_1

$$I_1 = I \times \dfrac{R_2}{R_1 + R_2}$$

Similarly, $I_2 = I \times \dfrac{R_1}{R_1 + R_2}$

Example 4.5

Two cables are connected in parallel, the resistances of which are 0·1 Ω and 0·06 Ω respectively. If the feed current is 80 A, calculate the current flowing through each cable.

Calling the current through the 0·1 Ω resistor I_1 and the current through the 0·06 Ω resistor I_2, then

$$I_1 = I \times \frac{R_2}{R_1 + R_2} = 80 \times \frac{0 \cdot 06}{0 \cdot 1 + 0 \cdot 06} = \frac{80 \times 0 \cdot 06}{0 \cdot 16} = \underline{30\ A}$$

$$I_2 = I \times \frac{R_1}{R_1 + R_2} = 80 \times \frac{0 \cdot 1}{0 \cdot 16} = \underline{50\ A}$$

As a check $I = I_1 + I_2$

i.e. $80\ A = 30\ A + 50\ A$

4.5. Temperature Coefficient

The resistance of most metallic conductors *increases* when subjected to a temperature rise. The reason is that when the material

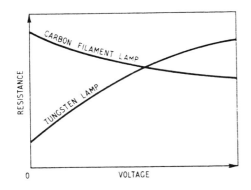

Fig. 4.9. Variation of resistance with voltage of tungsten and carbon-filament lamps.

is heated, the molecules of the substance are set into vibration. The movement of the electrons through the conductor is thus impeded, so that the electrical resistance becomes proportional to the temperature.

Carbon is a conductor which is said to possess a *negative temperature coefficient*, i.e. its resistance reduces as its temperature rises. Gases, electrolytes, semiconductors and insulators also follow this rule. Heat should be kept away from insulating materials, wherever possible, since it may be responsible for leakage currents and short-circuits.

An interesting experiment, which can be carried out in the technical college workshop or laboratory, is to place an ammeter in a 240-V lamp circuit and to note the current changes for varying voltages. The resulting resistance values can then be plotted against voltage. Figure 4.9 shows the curves obtained from such an experiment using a carbon-filament lamp and a tungsten lamp. From the curves it can be seen that the resistance of the tungsten filament increases with temperature, while that of the carbon lamp has a negative temperature coefficient.

Use of the temperature coefficient (symbol α) is a means of obtaining the actual resistance change. Temperature coefficient is normally defined as the resistance change (in ohms) per degree Celsius (Centigrade) change of a 1 Ω resistor, i.e. the rate of change of resistance with temperature, starting at 0°C.

For normal changes of temperature encountered the resistance change can be taken as uniform. The temperature coefficient (α) of copper is approximately 0·004 $\Omega/\Omega/°C$; thus from a resistance of 1 Ω at 0°C the resistance of the copper at 15°C = 1 + 15 × 0·004 = 1 + 0·06 = 1·06 Ω.

From the definition, the resistance at any other temperatures may be obtained.

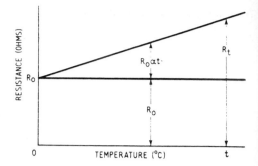

Fig. 4.10. Variation of resistance with temperature.

Resistance at temperature t = Resistance at $0°C$ + resistance increase

i.e.
$$R_t = R_0 + R_0 \alpha t$$
$$= R_0(1 + \alpha t)$$

Temperature coefficient may also be shown in a graphical form (Fig. 4.10).

Calculation involving temperature coefficient provides a simple method for obtaining the temperature at the interior of windings. The resistance at $0°C$ (R_0) may be difficult to obtain. By employing two equations this zero temperature may be eliminated from the calculation.

Calling R_1 the initial resistance at temperature t_1 and R_2 the final resistance at temperature t_2

then by division
$$\frac{R_1}{R_2} = \frac{R_0(1 + \alpha t_1)}{R_0(1 + \alpha t_2)}$$

Cancelling out R_0
$$\frac{R_1}{R_2} = \frac{1 + \alpha t_1}{1 + \alpha t_2}$$

Example 4.6

The resistance of the shunt winding of a d.c. machine is measured before and after a run of several hours.

The average values are 55 ohms and 63 ohms. Calculate the rise in temperature of the winding.

(Temperature coefficient of resistance of copper is 0·00428 ohm per ohm per °C.) [C]

$$\frac{R_1}{R_2} = \frac{1 + \alpha t_1}{1 + \alpha t_2}$$

Assuming a room temperature of 15°C

$$\frac{55}{63} = \frac{1 + 0.00428 \times 15}{1 + 0.00428 t_2}$$

$\therefore$ $0.8732(1 + 0.00428 t_2) = 1 + 0.0642$

$0.8732 + 0.003737 t_2 = 1.0642$

$0.003737 t_2 = 0.191$

$\therefore$ $t_2 = \dfrac{0.191}{0.003737}$ $= 51°C$

$\therefore$ Temperature rise $= 51°C - 15°C$ $= \underline{36°C}$

Example 4.7

A piece of resistance wire, 15·6 m long and of cross-sectional area 12 mm² at a temperature of 0°C, passes a current of 7·9 A when connected to a d.c. supply at 240 V. Calculate (a) resistivity of the wire, (b) the current which will flow when the temperature rises to 55°C. (The temperature coefficient of the resistance wire is 0·00029 Ω/Ω/°C.)
[C & G]

(a) $R = \dfrac{V}{I} = \dfrac{240}{7.9}$ $= 30.38\ \Omega$

But $R = \dfrac{\rho \times l}{10^6 \times a}$

where

ρ = resistivity in microhm-metre ($\mu\Omega$ m)
l = length in m
a = cross-sectional area in m²

$\therefore$ $\rho = \dfrac{R \times a \times 10^6}{l} = \dfrac{30.38 \times 12 \times 10^6}{15.6 \times 10^6} = \underline{23.37\ \mu\Omega\ m}$

(b) $R_t = R_0(1 + \alpha t)$

$= 30.38(1 + 0.00029 \times 55) = 30.38 \times 1.01595 = 30.86\ \Omega$

$I = \dfrac{V}{R} = \dfrac{240}{30.86}$ $= \underline{7.78\ A}$

4.6. Kirchhoff's Laws

These are considered as an extension of Ohm's Law; being self-evident they are extremely easy to understand. The reader has almost certainly applied the laws without calling them by name.

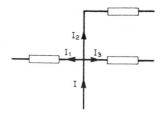

Fig. 4.11. Kirchhoff's current law.

(1) *Current Law. When currents meet at a junction, the sum of the currents entering the junction is equal to the sum leaving (Fig. 4.11).*

$$I = I_1 + I_2 + I_3$$

is obvious, but note by transposition

$$I_1 = I - (I_2 + I_3)$$
$$I_2 = I - (I_1 + I_3)$$
$$I_3 = I - (I_1 + I_2)$$

also
$$0 = I_1 + I_2 + I_3 - I$$

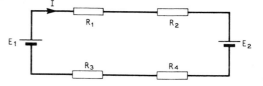

Fig. 4.12. Kirchhoff's voltage law where E_1 is greater than E_2.

(2) *Voltage Law. In any CLOSED circuit, the sum of the potential differences (i.e. the products of the current and resistances) are equal to the sum of the e.m.f.s.*

Assuming two e.m.f.s (Fig. 4.12) and that E_1 is greater than E_2 then,

$$E_1 - E_2 = IR_1 + IR_2 + IR_3 + IR_4$$
$$= I(R_1 + R_2 + R_3 + R_4)$$

Correct current directions are essential but the convention adopted for convenience when beginning a calculation is to assume that the currents flow round the circuit in a clockwise direction. Current I in the diagram, for example, is an assumed direction which will have to be reversed if the calculation shows it to be negative.

Example 4.8

(a) *Two batteries, one with an e.m.f. of 100 V and an internal resistance of 1·5 Ω and the other an e.m.f. of 60 V and an internal resistance of 0·5 Ω are connected in parallel to supply a load of 2 Ω. Calculate the current taken from each battery and the current flowing in the load.*

(b) *If the load resistance is increased to 3 Ω find the current now flowing.*   [C]

(a) The circuits may be drawn as follows, with assumed currents I_1 and I_2 (Fig. 4.13).

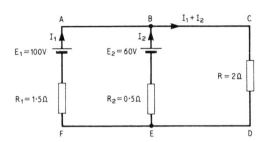

Fig. 4.13. Batteries connected in parallel.

The current $I_1 + I_2$ through the load resistance R follows from Kirchhoff's current law. Since these are two unknown quantities, two simultaneous equations are required for their solution. These equations are obtained from Kirchhoff's voltage law:

MESH ABEF

$$I_1 R_1 - I_2 R_2 = E_1 - E_2$$
$$1·5I_1 - 0·5I_2 = 100 - 60$$
$$\therefore \qquad 1·5I_1 - 0·5I_2 = 40 \qquad\qquad (1)$$

MESH ABCDEF

$$I_1 R_1 + (I_1 + I_2)R = E_1$$
$$1·5I_1 + 2I_1 + 2I_2 = 100$$
$$\therefore \qquad 3·5I_1 + 2I_2 = 100 \qquad\qquad (2)$$

(1) × 4 $$6\ I_1 - 2I_2 = 160 \qquad\qquad (3)$$
(2) + (3) $$9·5I_1 \qquad\quad = 260$$

$$\therefore \qquad\qquad I_1 = \frac{260}{9·5} \qquad\qquad = \underline{27·4\ A}$$

Substituting in (1)

$$1{\cdot}5 \times 27{\cdot}4 - 0{\cdot}5I_2 = 40$$

$$\therefore \quad 41{\cdot}1 - 40 = 0{\cdot}5I_2$$

$$I_2 = \frac{1{\cdot}1}{0{\cdot}5} \qquad = \underline{2{\cdot}2\text{ A}}$$

$$\text{current in load} = 27{\cdot}4\text{ A} + 2{\cdot}2\text{ A} \qquad = \underline{29{\cdot}6\text{ A}}$$

(b) MESH ABEF

As before

$$1{\cdot}5I_1 - 0{\cdot}5I_2 = 40 \tag{1}$$

MESH ABCDEF

$$1{\cdot}5I_1 + 3(I_1 + I_2) = 100$$

$$\therefore \quad 4{\cdot}5I_1 + 3I_2 = 100 \tag{2}$$

(1) × 6

$$9\ I_1 - 3I_2 = 240 \tag{3}$$

(2) + (3)

$$13{\cdot}5I_1 \qquad = 340$$

$$\therefore \qquad I_1 = \frac{340}{13{\cdot}5} \qquad = \underline{25{\cdot}2\text{ A}}$$

Substituting in (1)

$$1{\cdot}5 \times 25{\cdot}2 - 0{\cdot}5I_2 = 40$$

$$\therefore \qquad I_2 = \frac{40 - (1{\cdot}5 \times 25{\cdot}2)}{-0{\cdot}5} \qquad = \underline{-4{\cdot}4\text{ A}}$$

The minus sign signifies current flow in the reverse direction, i.e. from B to E.

$$\text{new current in load} = 25{\cdot}2 + (-4{\cdot}4) \qquad = \underline{20{\cdot}8\text{ A}}$$

4.7. Ring Main Calculations

Kirchhoff's laws are here particularly suitable for obtaining solutions.

Example 4.9

A 2-wire ring main is 600 m long, and is supplied at a point P with direct current at 240 V. A load of 70 A is taken from the main at point Q 150 m from P in one direction, and a further load of 50 A is taken at point R, 300 *m from* P *in the other direction. The resistance of the main is* 0·4 Ω/1000 *m of single core.*

Find the currents in value and direction in each section of the ring main, and calculate the voltages at points Q *and* R.   [C]

The ring main may be represented by Fig. 4.14.

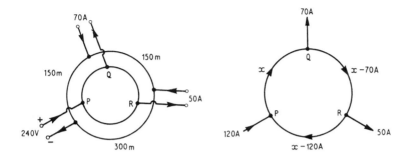

Fig. 4.14. Ring main representation.   Fig. 4.15. Simplified diagram of ring
                                          main.

resistance of one main, P to Q $= \frac{150}{1000} \times 0·4 = 0·06\ \Omega$

resistance of one main, Q to R $= \frac{150}{1000} \times 0·4 = 0·06\ \Omega$

resistance of one main, R to P $= \frac{300}{1000} \times 0·4 = 0·12\ \Omega$

By taking the current in section PQ as x, the value in the other sections may be obtained in terms of x from Kirchhoff's first law, as noted in the simplified diagram (Fig. 4.15). Clockwise directions are taken but if the results produce negative values the directions will have to be reversed to those as shown. Attention is also drawn to the fact that there is no source of voltage in the ring itself, i.e. the ring e.m.f. is zero.

From Kirchhoff's second law and working round the ring starting from P :

$$0·06x + 0·06\,(x - 70) + 0·12\,(x - 120) = 0$$

∴ $$0·06x + 0·06x - 4·2 + 0·12x - 14·4 = 0$$

∴ $$0·24x = 18·6$$

$$x = \frac{18·6}{0·24} \quad = 77·5\ \text{A}$$

∴   Current in section P–Q $= \underline{77\cdot5\,A}$

Current in section Q–R $= 77\cdot5 - 70$ $= \underline{7\cdot5\,A}$

Current in section R–P $= 77\cdot5 - 120$ $= \underline{-42.5\,A}$

Thus the current in this last section flows in the reverse direction to that shown in Fig. 4.15.

Voltage drop P–Q $= 77\cdot5 \times 0\cdot12$ (+ ve and − ve main) $= \underline{9\cdot3\,V}$

∴

voltage at point Q $= 240 - 9\cdot3$ $= \underline{230\cdot7\,V}$

Voltage drop P–R $= 42\cdot5 \times 0\cdot24$ $= \underline{10\cdot2\,V}$

∴

voltage at point $R = 240 - 10\cdot2$ $= \underline{229\cdot8\,V}$

Alternatively, the voltage at R could be found by calculating the volt drop Q–R, i.e. $7\cdot5 \times 0\cdot12 = 0\cdot9$ V, and deducting this figure from the actual voltage at Q, i.e. $230\cdot7$ V, this subtraction, $230\cdot7$ V $- 0\cdot9$ V, giving the voltage at point R of $229\cdot8$ V.

Applying the calculation technique to the domestic ring circuit is of interest and gives useful practice.

Example 4.10

Calculate the current in each section of the ring circuit as illustrated in Fig. 4.16.

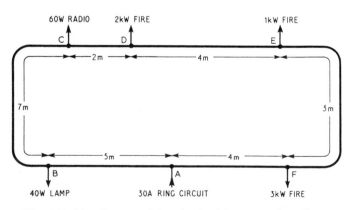

Fig. 4.16. Line diagram of ring circuit with connected appliances.

For a standard voltage of 240 V and applying the formula $I = P/V$

40 W lamp	40/240	=	0·167
60 W radio	60/240	=	0·25
2 kW fire	2000/240	=	8·333
1 kW fire	1000/240	=	4·167
3 kW fire	3000/240	=	12·5
		Total	25·417 A

Calling the current from A to B, x, and applying Kirchhoff's first law the current through the various sections can be stated in terms of x (Fig. 4.17(a) and 4.17(b)):

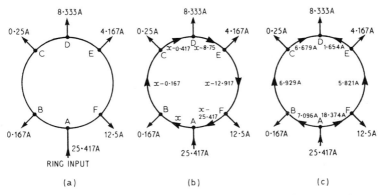

Fig. 4.17. (a) Ring circuit with currents flowing to various appliances. (b) Current through the various section in terms of x. (c) Final current values and corrected directions of current flow.

A to B		= x
B to C		= $x - 0·167$
C to D = $x - 0·167 - 0·25$		= $x - 0·417$
D to E = $x - 0·417 - 8.333$		= $x - 8·75$
E to F = $x - 8·75 - 4·167$		= $x - 12·917$
F to A = $x - 12·917 - 12·5$		= $x - 25·417$

From first principles, $R = \rho \dfrac{l}{a}$

Resistance of 100 m of 2·5 mm² cable

$$= \frac{0\cdot017 \times 100 \times 10^6}{10^6 \times 2\cdot5} = 0\cdot68 \ \Omega$$

Resistance of section AB $= 0\cdot68 \times \frac{5}{100}$ $= 0\cdot034 \ \Omega$

BC $= 0\cdot68 \times \frac{7}{100}$ $= 0\cdot0476 \ \Omega$

CD $= 0\cdot68 \times \frac{2}{100}$ $= 0\cdot0136 \ \Omega$

DE $= 0\cdot68 \times \frac{4}{100}$ $= 0\cdot0272 \ \Omega$

EF $= 0\cdot68 \times \frac{3}{100}$ $= 0\cdot0204 \ \Omega$

FA $= 0\cdot68 \times \frac{4}{100}$ $= 0\cdot0272 \ \Omega$

Applying Kirchhoff's second law, the products of the current and resistance of each section are thus:

AB $= 0\cdot034x$

BC $= (x-0\cdot167)0\cdot0476$ $= 0\cdot0476x-0\cdot0079$

CD $= (x-0\cdot417)0\cdot0136$ $= 0\cdot0136x-0\cdot0056$

DE $= (x-8\cdot75)0\cdot0272$ $= 0\cdot0272x-0\cdot238$

EF $= (x-12\cdot917)0\cdot0204$ $= 0\cdot0204x-0\cdot2635$

FA $= (x-25\cdot417)0\cdot0272$ $= 0\cdot0272x-0\cdot6913$

Total $= 0\cdot17x-1\cdot2063$

i.e. $0 = 0\cdot17x-1\cdot2063$

$$0\cdot17x = 1\cdot2063$$

$$x = \frac{1\cdot2063}{0\cdot17 \ A}$$

$$= 7\cdot096 \ A$$

The value of the current in each section is obtained by substituting 7·096 A for x. This gives:

AB $= 7\cdot096$ A

BC $7\cdot096-0\cdot167$ $= 6\cdot929$ A

CD $7\cdot096-0\cdot417$ $= 6\cdot679$ A

DE $7\cdot096-8\cdot75$ $= -1\cdot654$ A

EF $7\cdot096-12\cdot917$ $= -5\cdot821$ A

FA $7\cdot096-25\cdot417$ $= -18\cdot374$ A

Inserting these figures, the final line diagram of the currents, with corrected directions, is set out in Fig. 4.17(c).

The example has been worked out in full. However, since the resistance of a conductor is proportional to its length, the calculations could be much simplified by using the actual distances instead of working out the various resistances.

4.8. Loads Supplied at Both Ends

For parallel loads fed in both directions by supplies of equal voltage, the calculations for determining the value of the currents in each cable section are exactly the same as for a single ring main. Where the feeds are of different voltage, the net e.m.f. will not be zero but the difference between these two voltages, the effect being the same for batteries with different voltages feeding into a common load.

EXERCISES

1. Write down Kirchhoff's laws and use them in working the following calculation:

A 2-wire ring main is 600 m long, and is supplied at a point P with direct current at 240 V. A load of 70 A is taken from the main at point Q, 150 m from P in one direction, and a further load of 50 A is taken at a point R, 300 m from P in the other direction. The resistance of the main is 0·4 Ω/1000 m of single core.

Find the currents in value and direction in each section of the ring main, and calculate the voltages at points Q and R.   [C]

2. Four resistance coils, AB, BC, CD and DA, are connected together in order, to form a square ABCD.

A d.c. supply, comprising 10 cells each of e.m.f. 2·1 volts and internal resistance 0·15 ohm, is connected across A and C by wires of total resistance 0·3 ohm, so that current flows in direction A to C. A high-resistance voltmeter taking negligible current is connected between B and D.

If the respective values of the resistances are, AB = 2 ohms, BC = 7 ohms, CD = 3 ohms and DA = 6 ohms, find the reading of the voltmeter in value and direction.

What change in the value of AB is necessary for the voltmeter reading to be zero?   [C]

3. Write down Kirchhoff's laws and use them in working the following calculation. Four resistances are connected in a closed circuit in the form of a square ABCD, where AB = 20 Ω, BC = 30 Ω, CD = 40 Ω and DA = 50 Ω. A galvanometer of resistance 80 Ω is connected across B and D, and a cell of e.m.f. 2 V and negligible resistance is connected across A and C. Find the value of the current in the galvanometer, and show clearly its direction.   [C]

4. A 150 Ω resistance coil AB is connected to terminals at 240 V d.c. Calculate the value of a further resistance coil which when connected to the mid-point of AB and the end A, will carry a current of 0·8 A.   [C]

5. A resistance coil AB of 100 Ω resistance is to be used as a potentiometer, and is connected to a d.c. supply at 230 V. Find by calculation the position of a tapping point C between A and B such that a current of 2 A will flow in a resistance of 50 Ω connected across A and C.   [C]

6. When working normally the temperature of the filament in a 230-V, 150-W gas-filled, tungsten lamp is 2750°C. Assuming a room temperature of 16°C, calculate: (a) the normal current taken by the lamp, (b) the current taken at the moment of switching on. (Temperature coefficient of tungsten = 0·0047 Ω/Ω/°C at 0°C).   [C]

7. Describe with a diagram of connexions one form of direct-reading commercial potentiometer. Show how this instrument could be used to calibrate an ammeter of range 0–20 A. Produce a diagram of the auxiliary circuit used.

8. Describe the principle of the d.c. potentiometer. Explain clearly, with the aid of a circuit diagram, how the value of an unknown resistance may be obtained by comparing it with a standard resistance by means of the potentiometer.

9. Give reasons for the increase in temperature in different parts of a d.c. generator on load. The shunt winding of a d.c. generator has a resistance of 165 Ω at a temperature of 15°C. Calculate the resistance when the temperature has reached 47°C. The temperature coefficient of copper may be taken as 0·00428 Ω/Ω/°C.

Chapter 5

A.C. CIRCUITS

5.1. A.C. Series and Parallel Circuits

Explanations are given by worked examples based on earlier study.

Example 5.1

A capacitor of capacitance C farads, an inductance of L henrys and a resistance of R ohms are connected in series. A single-phase supply of V volts at a frequency of f hertz is applied to the ends of the circuit to pass a current of I amperes.

(a) Write down the formula giving I in terms of the other quantities, and explain with the aid of a phasor diagram how the formula is derived.

(b) Find the value of V if the other values are: C = 60 microfarads, L = 0·2 H, R = 5 Ω, f = 50 Hz, I = 20 A. [C]

(a) $I = \dfrac{V}{Z}$ where Z, impedance

$$= \sqrt{[R^2 + (X_L - X_C)^2]}$$

X_L, inductive reactance

$$= 2\pi f L$$

$$= \frac{V}{\sqrt{[R^2 + (X_L - X_C)^2]}}$$

X_C, capacitive reactance

$$= \frac{10^6}{2\pi f C}$$

The voltages across R, L and C (Fig. 5.1(a)) are together phasorially equal to V as shown by the voltage phasor diagram.

From the phasor diagram it can be seen that the reference current I lags V_L by 90°, leads V_C by 90° and lags or leads the supply voltage V by some angle ϕ. This latter angle depends upon the relative magnitudes of V_L and V_C. If V_C is numerically greater than V_L, the current

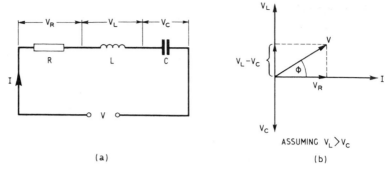

(a)

(b)

Fig. 5.1. In an a.c. series circuit, the individual voltage drops must be added phasorially to give the supply voltage.

will lead the supply voltage, also

$$V_R = \text{current} \times \text{resistance} \qquad = IR$$

$$V_L = \text{current} \times \text{inductive reactance} \qquad = IX_L$$

$$V_C = \text{current} \times \text{capacitive reactance} \qquad = IX_C$$

$$V = \text{current} \times \text{impedance} \qquad = IZ$$

If each phasor of Fig. 5.1(b) is divided by I, the magnitude will be reduced but the general shape of the diagram will be retained and the angle ϕ does not change. The voltage phasor diagram can therefore be simplified to the general form of an impedance triangle (Fig. 5.2).

By Pythagoras' Theorem, $Z = \sqrt{[R^2 + (X_L - X_C)^2]}$. X_C may be numerically more than X_L in which case $(X_L - X_C)$ will be negative, but squaring will produce a positive value.

$$(b) \qquad Z = \sqrt{\left[R^2 + \left(2\pi f L - \frac{10^6}{2\pi f C} \right)^2 \right]}$$

$$= \sqrt{\left[5^2 + \left(2\pi \times 50 \times 0.2 - \frac{10^6}{2\pi \times 50 \times 60} \right)^2 \right]} \qquad = 10.95\ \Omega$$

$$V = IZ = 20 \times 10.95 \qquad = \underline{219\ V}$$

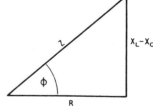

Fig. 5.2. General form of impedance triangle.

Example 5.2

A resistance of 12 Ω and a capacitance of 300 µF are connected in series. A coil of inductance 0·5 henry and resistance 8 Ω is in parallel with these. A single-phase supply of 240 V 50 Hz is applied to the ends of the combination.

Determine by calculation or graphically, (a) the current and power factor in each of the parallel circuits and (b) the total current from the supply and the power factor. [C]

(a) The circuit is drawn in Fig. 5.3(a)

$$Z = \sqrt{\left[R^2 + \left(\frac{10^6}{2\pi f C} \right)^2 \right]} \quad \text{where } R \text{ and } C \text{ are in series}$$

$$Z_1 = \sqrt{\left[12^2 + \left(\frac{10^6}{2\pi \times 50 \times 300} \right)^2 \right]} = \sqrt{256 \cdot 6} \qquad = 16 \,\Omega$$

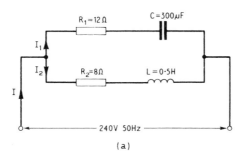

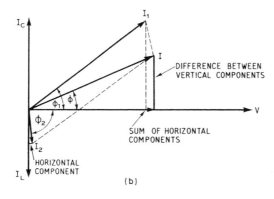

Fig. 5.3. A.C. parallel circuit.

$\therefore$

$$I_1 = \frac{240}{16} \qquad\qquad\qquad = 15\,A$$

Referring to Fig. 5.3(b), in which voltage is taken as the datum line.

Power factor $\qquad \cos\phi_1 = \dfrac{R_1}{Z_1} = \dfrac{12}{16} \qquad = 0.75\ \text{leading}$

$Z = \sqrt{[R^2 + (2\pi f L)^2]} \quad$ where R and L are in series

$\therefore$

$Z_2 = \sqrt{[8^2 + (2\pi \times 50 \times 0.5)^2]} = \sqrt{(24\,744)} \qquad = 157.3\ \Omega$

$\therefore$

$$I_2 = \frac{240}{157.3} \qquad\qquad\qquad = 1.526\,A$$

Power factor, $\qquad \cos\phi_2 = \dfrac{R}{Z_2} = \dfrac{8}{157.2} \qquad = 0.05\ \text{lagging}$

(b) In branch 1, $\cos\phi_1 = 0.75$

$\therefore \qquad\qquad\qquad \phi_1 = 41°\,24'$ and $\sin\phi_1 \qquad = 0.6613$

In branch 2, $\qquad \cos\phi_2 = 0.05$

$\therefore \qquad\qquad\qquad \phi_2 = 87°\,4'$ and $\sin\phi_2 \qquad = 0.9987$

Again with reference to Fig. 5.3(b), I_1 and I_2 can be split into in phase (horizontal) and quadrature (vertical) components.

By applying Pythagoras' Theorem:

$I = \sqrt{[(\text{Sum of horizontal components})^2 + (\text{Difference vertical}}$
$$\text{components})^2]$$

$= \sqrt{[(I_1\cos\phi_1 + I_2\cos\phi_2)^2 + (I_1\sin\phi_1 - I_2\sin\phi_2)^2]}$

$= \sqrt{[(15 \times 0.75 + 1.526 \times 0.05)^2 + (15 \times 0.6613 - 1.526 \times 0.9987)^2]}$

$= \sqrt{[(11.25 + 0.076)^2 + (9.92 - 1.524)^2]} = \sqrt{(11.33^2 + 8.39^2)}$

$$= 14.1\,A$$

N.B. Differences between vertical components are taken because all values above the voltage reference line are positive and those below the line are negative.

$$\cos\phi = \frac{11.33}{14.1} \qquad\qquad = 0.8\ \text{(leading)}$$

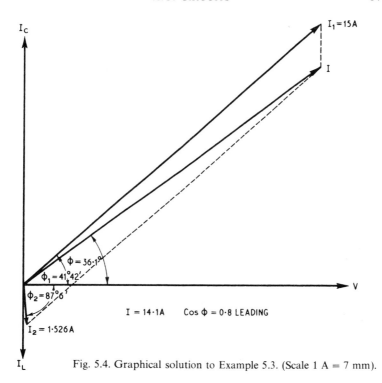

Fig. 5.4. Graphical solution to Example 5.3. (Scale 1 A = 7 mm).

The graphical (Fig. 5.4) solution is quicker since calculations are only required for the branch currents and phase angles. It clearly shows that I is the phasorial sum of I_1 and I_2.

Example 5.3

When supplied with current at 240 *V, single phase, at* 50 *Hz a certain inductive coil takes* 13·62 *A.*

If the frequency of supply is changed to 40 *Hz the current increases to* 16·12 *A.*

Calculate the resistance and inductance of the coil.   [C]

At 50 *Hz*

$$Z = \frac{V}{I}$$

$$= \frac{240}{13\cdot62} = 17\cdot62\,\Omega$$

Also $Z = \sqrt{[R^2 + (2\pi f L)^2]}$

$\therefore$ $17.62 = \sqrt{[R^2 + (314L)^2]}$ Squaring both sides

$310.5 = R^2 + (314L)^2$ (1)

At 40 Hz $Z = \dfrac{240}{16.12} = 14.89 \, \Omega$

$\therefore$ $14.89 = \sqrt{[R^2 + (251L)^2]}$ Squaring both sides

$221.6 = R^2 + (251L)^2$ (2)

Subtracting (2) from (1)

$88.9 = (314L)^2 - (251L)^2$

$ = (314L + 251L)(314L - 251L)$ (difference two squares)

$88.9 = (565 \times 63)L^2$

$\therefore \; L = \sqrt{\dfrac{88.9}{565 \times 63}}$ $= \underline{0.05 \text{ H}}$

Substituting in (1)

$310.5 = R^2 + (314 \times 0.05)^2$

$R^2 = 310.5 - 15.2$

$R = \sqrt{295.3}$ $= \underline{17.2 \, \Omega}$

(*Comments.* This is one of those questions which will enable students with sufficient practice of elementary mathematics to spot the simultaneous equations involved and to determine the solution with the minimum of effort.)

Example 5.4

A coil of insulated wire of resistance 8 ohms and inductance 0.03 henry is connected to an a.c. supply at 240 V, 50 Hz.
   Calculate:
   (a) The current, the power, and the power factor,
   (b) The value (in microfarads) of a capacitance which, when connected in series with the above coil, causes no change in the values of current and power taken from the supply.   [C]

(a) $Z = \sqrt{(R^2 + X_L^2)}$

$Z = \sqrt{[R^2 + (2\pi f L)^2]}$

$= \sqrt{[8^2 + (2\pi \times 50 \times 0.03)^2]} = \sqrt{(8^2 + 9.434^2)}$

$= \sqrt{(64 + 89)}$ $= 12.37\ \Omega$

Current, $I = \dfrac{V}{Z} = \dfrac{240}{12.37}$ $= \underline{19.4\ A}$

Power $= I^2 R = 19.4^2 \times 8$ $= \underline{3.012\ kW}$

Power factor $= \dfrac{R}{Z} = \dfrac{8}{12.37}$ $= \underline{0.65\ lagging}$

(Power could also be obtained by means of the formula $P = VI \cos\phi$.)

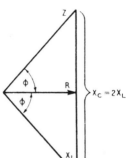

Fig. 5.5. With R and L in series the same value of impedance can be maintained by a capacitor producing a capacitive reactance equal to twice the inductive reactance.

(b) To maintain the same current and power, the impedance of the circuit has to be unchanged. Thus the effect of placing the capacitor in series must be to cause the current to lead by the same angle as it previously lagged. This may be seen by means of the phasor diagram (Fig. 5.5).

$$X_C = 2X_L$$

$$= 2 \times 9.434$$

$$= 18.868\ \Omega$$

But $$X_C = \dfrac{10^6}{2\pi f C}$$

∴ $$18.868 = \dfrac{3183}{C}\left(\text{since } \dfrac{10^6}{314} = 3183\right)$$

∴ $$C = \dfrac{3183}{18.867}$$ $= \underline{168.7\ \mu F}$

5.2. Resonance

Inductive reactance $X_L = 2\pi f L$, thus X_L is proportional to the frequency f. Also capacitive reactance $X_C = \dfrac{1}{2\pi f C}$ so that X_C is inversely proportional to frequency. If therefore both inductance and capacitance are present in a circuit and the frequency is varied in regular steps to allow the inductive and capacitive reactance to be calculated at each stage, an examination of the results will show that there is a point where $X_L = X_C$. This can also be seen by plotting the points on a graph as shown in Fig. 5.6.

The circuit has a minimum impedance at the 'crossover point'. At this position, since X_L and X_C are equal, but opposite in phase, they neutralise each other *and the impedance then becomes numerically equal to the resistance*. It also follows that the current is at its maximum and the circuit is now said to be in a state of resonance.

The frequency (f_0) at which resonance occurs can be determined by equating the two reactances:

$$2\pi f L = \frac{1}{2\pi f C}$$

transposing

$$f^2 = \frac{1}{4\pi^2 LC}$$

$\therefore$

$$f_0 = \frac{1}{2\pi\sqrt{(LC)}}$$

where L is in henrys and C in farads.

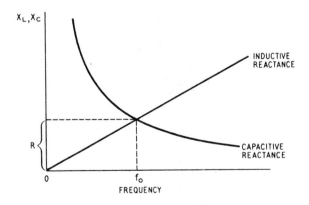

Fig. 5.6. Resonance conditions.

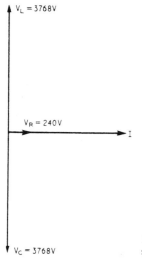

$V_L = 3768V$

$V_R = 240V$

$V_C = 3768V$

Fig. 5.7. Phasor diagram of resonant conditions in series circuit.

Example 5.5

A resistor of 6 Ω *is connected in series with an inductor of* 0·25 *H and a capacitor of* 28·14 μF *on a* 240 V *circuit, calculate*
(a) *resonant frequency*;
(b) *the voltage across each component*;
(c) *comment briefly on the resultant phasor diagram.*

(a) Resonant frequency,

$$f_0 = \frac{1}{2\pi\sqrt{(LC)}} = \frac{1}{2\pi\sqrt{\left(\dfrac{0·25 \times 28·14}{10^6}\right)}}$$

$$= \frac{1}{2\pi\sqrt{7·035 \times 10^{-6}}} \qquad = \underline{60\ \text{Hz}}$$

(b) $\qquad X_L = 2\pi f L = 2\pi \times 60 \times 0·25 \qquad = \underline{94·2\ \Omega}$

$$X_C = \frac{10^6}{2\pi f_0 C} = \frac{10^6}{2\pi \times 60 \times 28·14} \qquad = \underline{94·2\ \Omega}$$

But $\qquad Z = \sqrt{[R^2 + (X_L - X_C)^2]}$

$$= R \text{ since } X_L = X_C$$

∴   Current $\qquad I = \dfrac{V}{R} = \dfrac{240}{6} \qquad = \underline{40\ \text{A}}$

$$\text{Voltage across resistor} = IR = 40 \times 6 \qquad\qquad = \underline{240 \text{ V}}$$

$$\text{Voltage across inductor} = IX_L = 40 \times 94{\cdot}2 \qquad = \underline{3768 \text{ V}}$$

$$\text{Voltage across capacitor} = IX_c = 40 \times 94{\cdot}2 \qquad = \underline{3768 \text{ V}}$$

(c) The phasor diagram (Fig. 5.7) illustrates the series resonant circuit. It shows that dangerously high voltages can arise across the reactive components.

On the other hand, advantage can be taken of series resonance to achieve quick starting in some types of fluorescent lamp circuit; the correct choice of the reactance values ensures a high striking voltage.

5.3. Current Resonance

Current resonance occurs in parallel circuits at resonant frequency. Under these conditions abnormally high currents flow in each branch.

5.4. 3-phase Star-connected System

Sometimes known as the Y connected system, the method of connexions is given in Fig. 5.8, as feeding a balanced resistive load.

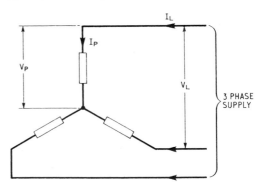

Fig. 5.8. Three resistors, star-connected.

For the sake of clarity, line (V_L, I_L) and phase (V_P, I_P) notation are only shown for one phase. Clearly $I_L = I_P$, i.e. for *star connexions the line current is equal to the phase current*. The phasor diagram (Fig. 5.9) serves as an aid in determining the relation between the line and phase voltages. The voltage between two lines is the phasor *difference* of two phase voltages. Therefore one of the phase voltage phasors is reversed to produce minus V_P (i.e. $-V_P$). From the geometry of the figure

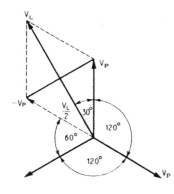

Fig. 5.9. Voltage phasor diagram of 3-phase star-connected system.

$$\frac{\frac{1}{2}V_L}{V_P} = \cos 30°$$

$$\therefore \quad \tfrac{1}{2}V_L = V_P \times \tfrac{1}{2}\sqrt{3}$$

$$\therefore \quad V_L = \sqrt{3}\, V_P$$

The line voltage is $\sqrt{3}$ times the phase voltage ($\sqrt{3} = 1·73$).
For a balanced load the general formula for 3-phase power becomes:

$$P = \sqrt{3}\, VI \cos \phi, \quad \text{where } V \text{ and } I \text{ are line values}$$

It is left as an exercise for the student to check that for a balanced load the total power is also equal to three times the power developed in each phase.

5.5. 3-phase Delta-connected System

Also called the mesh connexion, the general scheme can be seen in Fig. 5.10. Again for simplicity only one set of phase and line values are marked.

Here $V_L = V_P$, i.e. *for delta connexion the line voltage is equal to the phase voltage.* From the phasor diagram (Fig. 5.11), the line current is equal to the phasor difference of two phase currents

$$\frac{\frac{1}{2}I_L}{I_P} = \cos 30°$$

$$\therefore \quad \tfrac{1}{2}I_L = I_P \times \tfrac{1}{2}\sqrt{3}$$

$$\therefore \quad I_L = \sqrt{3}\, I_P$$

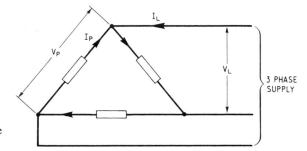

Fig. 5.10. 3-phase
delta connexions.

The line current is $\sqrt{3}$ times the phase current. As before the
general formula for 3-phase power becomes:

$$P = \sqrt{3}\, VI \cos \phi, \quad \text{where } V \text{ and } I \text{ are line values.}$$

Also as for star connexion, the total power can be obtained by three
times the power in each phase.

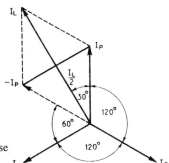

Fig. 5.11. Current phasor diagram for 3-phase
delta-connected system.

Example 5.6

*A 3-phase load consists of three similar inductive coils, each of
resistance 50 Ω and inductance 0·3 H. The supply is 415 V, 50 Hz.
Calculate: (a) the line current, (b) the power factor, and (c) the total
power when the load is (i) star-connected, and (ii) delta-connected.*

[C]

Impedance of each coil,

$$Z_P = \sqrt{[R^2 + (2\pi f L)^2]}$$
$$= \sqrt{[50^2 + (2\pi \times 50 \times 0.3)^2]} = 106.6 \ \Omega$$

(i) STAR-CONNECTED

(a) Line voltage is equal to $\sqrt{3}$ times the phase voltage

$\therefore$
$$V_L = \sqrt{3} \, V_P$$

$\therefore$
$$V_P = \frac{V_L}{\sqrt{3}} = \frac{415}{\sqrt{3}} \qquad = 240 \text{ V}$$

$$\text{Current in each phase} = \frac{\text{phase voltage}}{\text{impedance}} = \frac{240}{106 \cdot 6} = 2 \cdot 25 \text{ A}$$

$\therefore$
$$\text{Line current} = \underline{2 \cdot 25 \text{ A}}$$

(b)
$$\text{Power factor} = \frac{R}{Z_P} = \frac{50}{106 \cdot 6} \qquad = \underline{0 \cdot 47 \text{ lagging}}$$

(c)
$$\text{Power} = \sqrt{3} \, VI \cos \phi$$
$$= \sqrt{3} \times 415 \times 2 \cdot 25 \times 0 \cdot 47 = \underline{762 \text{ W}}$$

Alternatively,
$$\text{Power} = 3 \times \text{power in each phase}$$
$$= 3 \times V_P I_P \cos \phi$$
$$= 3 \times 240 \times 2 \cdot 25 \times 0 \cdot 47 \qquad = \underline{762 \text{ W}}$$

(ii) DELTA-CONNECTED

(a)
$$I_P = \frac{V_P}{Z_P} = \frac{415}{106 \cdot 6} \qquad = 3 \cdot 9 \text{ A}$$

$$I_L = \sqrt{3} \, I_P = \sqrt{3} \times 3 \cdot 9 \qquad = \underline{6 \cdot 75 \text{ A}}$$

(b)
$$\text{Power factor} = \frac{R}{Z_P} = \frac{50}{106 \cdot 6} \qquad = \underline{0 \cdot 47 \text{ lagging}}$$

(c)
$$\text{Power} = \sqrt{3} \, VI \cos \phi$$
$$= \sqrt{3} \times 415 \times 6 \cdot 75 \times 0 \cdot 47 = \underline{2280 \text{ W}}$$

Alternatively,
$$\text{Power} = 3 \times V_P I_P \cos \phi$$
$$= 3 \times 415 \times 3 \cdot 9 \times 0 \cdot 47 \qquad = \underline{2280 \text{ W}}$$

5.6. Unbalanced 3-phase Loads

The phase windings of a 3-phase motor are symmetrical so that there is no unbalance. This is also true for three equal resistive loads where balanced over three phases. However, differing single-phase loads if connected to a 3-phase supply create an unbalance, which requires a neutral wire connected to the star point to carry these out-of-balance currents.

Example 5.7

A 415-V 3-phase, 4-wire system supplies power to three noninductive loads. The loads are 25 kW between red and neutral, 30 kW between yellow and neutral and 12 kW between blue and neutral.

Calculate (a) the current in each line wire, and (b) the current in the neutral conductor.   [C]

Having calculated the current in each phase, the graphical solution gives the simplest method of obtaining the neutral current.

(a) $$\text{Phase voltage} = \frac{415}{\sqrt{3}} = 240 \text{ V}$$

∴ $$\text{Current in Red phase} = \frac{25\,000 \text{ W}}{240 \text{ V}} = \underline{104\cdot2 \text{ A}}$$

$$\text{Current in Yellow phase} = \frac{30\,000 \text{ W}}{240 \text{ V}} = \underline{125 \text{ A}}$$

$$\text{Current in Blue phase} = \frac{12\,000 \text{ W}}{240 \text{ V}} = \underline{50 \text{ A}}$$

(b) The currents in the red, yellow and blue phases have a phase difference of 120° and are set out to scale as in Fig. 5.12, a suggested

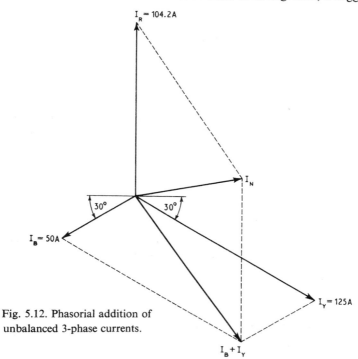

Fig. 5.12. Phasorial addition of unbalanced 3-phase currents.

scale being 10 mm = 1 A. Two of the phase currents, say blue and yellow are first combined phasorially. To obtain the current in the neutral, this composite current is then phasorially combined with the remaining phase current. Finally, the magnitude of the neutral current is obtained by measurement and then scaling back, thus $I_N = 67$ A.

By calculation. Convert each of the phase currents into horizontal and vertical components. The neutral current is then determined by applying Pythagoras' Theorem to the resultant components.

Resultant horizontal component

$$= I_Y \cos 30° - I_B \cos 30°$$

$$= 125 \times 0.866 - 50 \times 0.866$$

$$= 108.25 - 43.3 \qquad\qquad = 64.95 \text{ A}$$

Resultant vertical component

$$= I_R - I_Y \sin 30° - I_B \sin 30°$$

$$= 104.2 - 125 \times 0.5 - 50 \times 0.5$$

$$= 104.2 - 62.5 - 25 \qquad\qquad = 16.7 \text{ A}$$

Current in neutral conductor I_N

$$= \sqrt{(64.95^2 + 16.7^2)} \qquad\qquad = 67 \text{ A}$$

A useful exercise is to apply the method of phasorial addition to the *balanced* 3-phase circuit. It will show that the phasorial sum is zero. A further proof is provided by Kirchhoff's first law, 'when currents meet at a junction the algebraic sum is zero'. Here $I_R + I_Y + I_B = 0$.

5.7. Addition of Parallel Loads

The techniques adopted are of great practical usefulness. A simple single-phase case is considered first in order to illustrate the methods involved.

Example 5.8

(a) *What is meant by power factor?*

(b) *The installation in a factory carries the following load: lighting 50 kW, heating 30 kW and power 45 kW. Assuming that the lighting and heating loads are non-inductive and the power has an overall efficiency of 87% at a power factor of 0·7 lagging, calculate:*

  (i) *the total loading in kW,*
  (ii) *the kVA demand at full load.* **[B]**

(a) Consideration of the a.c. waveform shows that where a phase difference occurs, the product of the voltage and current does not give the power in watts but must, in addition, be multiplied by some other number less than unity, called the power factor,

i.e. $$P = VI \times \text{power factor (in watts)}$$

∴ $$\text{Power factor} = \frac{P\,(\text{watts})}{VI\,(\text{volt-amperes})}$$

For sinusoidal supplies, a convenient form is

$$\text{Power factor} = \cos \phi$$

where ϕ is the angle of lag *or lead* between V and I.

  (b) (i) Motor Load

$$\text{Efficiency} = \frac{45\ \text{kW}}{\text{input}}$$

$$\frac{87}{100} = \frac{45}{\text{input}}$$

∴ $$\text{Input} = \frac{45 \times 100}{87} = 51 \cdot 72\ \text{kW}$$

Lighting load $= 50\ \text{kW}$

Heating load $= 30\ \text{kW}$

Total loading $= 131 \cdot 72\ \text{kW}$

  (ii) $$\cos \phi = 0 \cdot 7$$

∴ $$\text{motor phase angle} = 45° \, 26'$$

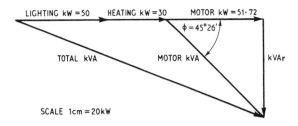

Fig. 5.13. Addition of simple parallel loads.

A phase diagram may now be drawn (Fig. 5.13)

From diagram   kVA demand at full load = 141 kVA

Mathematical calculation gives improved accuracy:

Reactive kilovolt-amperes $= 51 \cdot 72 \times \tan \phi$

$$= 51 \cdot 72 \times \tan 45° \, 26' = 52 \cdot 51 \text{ kVA}$$

Total kilovolt-amperes $= \sqrt{(131 \cdot 72^2 + 52 \cdot 51^2)} = 141 \cdot 8 \text{ kVA}$

Example 5.9

A single-phase supply is connected to the following loads:
(i) a 20-kW lighting load comprising of filament lamps;
(ii) an induction motor load of 75 kW at p.f. of 0·7 lagging;
(iii) a synchronous motor load of 45 kW at p.f. of 0·92 leading.
Draw a phasor diagram showing power and kVA and calculate
the kW, the kVA and power factor of the total load.

$$\cos \phi = \frac{kW}{kVA} \quad \therefore \quad kVA = \frac{kW}{\cos \phi}$$

With suffixes to indicate the various loads:

Lighting $\qquad\qquad kVA_1 = \dfrac{20}{1} \qquad\qquad = 20 \text{ kVA}$

Induction motor $\qquad kVA_2 = \dfrac{75}{0 \cdot 7} \qquad\qquad = 107 \cdot 2 \text{ kVA}$

Synchronous motor $\qquad kVA_3 = \dfrac{45}{0 \cdot 92} \qquad\qquad = 48 \cdot 91 \text{ kVA}$

$$\cos \phi_2 = 0 \cdot 7 \quad \therefore \quad \phi_2 = 45° \, 34'$$
$$\cos \phi_3 = 0 \cdot 92 \quad \therefore \quad \phi_3 = 23° \, 4'$$

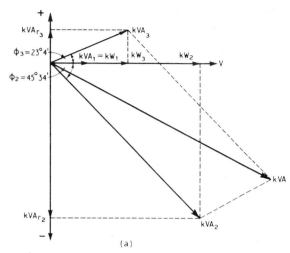

(a)

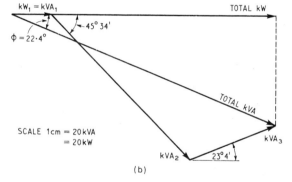

(b)

Fig. 5.14.
Summation of
lagging and
leading
parallel loads.

Insert these values in the phasor diagram (not drawn to scale), as observed in Fig. 5.14(a). As the loads are parallel, the supply voltage is given as the reference line. Values above this line are positive and are negative when below and refer to leading and lagging quantities respectively.

Reactive components:

$$\text{kVAr}_1 = \text{kVA}_1 \sin \phi_1 = 20 \times 0 \qquad\qquad = 0$$
$$\text{kVAr}_2 = \text{kVA}_2 \sin \phi_2 = 107 \cdot 2 \times 0 \cdot 7142 \qquad = -76 \cdot 56 \text{ kVAr}$$
$$\text{kVAr}_3 = \text{kVA}_3 \sin \phi_3 = 48 \cdot 91 \times 0 \cdot 3918 \qquad = \underline{19 \cdot 16 \text{ kVAr}}$$

$$\text{Total} \qquad\qquad\qquad\qquad\qquad\qquad = -57 \cdot 4 \text{ kVAr}$$

Total kilowatts

$$= 20 + 75 + 45 \qquad\qquad\qquad = 140 \text{ kW}$$

Total kilovolt-amperes

$$= \sqrt{(kW^2 + kVAr^2)}$$
$$= \sqrt{[140^2 + (-57\cdot4^2)]} \qquad\qquad = 151\cdot3 \text{ kVA}$$

Power factor of total load $= \dfrac{kW}{kVA} = \dfrac{140}{151\cdot3} = 0\cdot9253 \text{ lagging}$

Graphical Solution

As in the previous example the method is 'funicular' (carriages of a railway train), i.e. the phasors are continued along consecutively. Scaling back from Fig. 5.14(b)

$$\text{Total kilovolt-amperes} = 152 \text{ kVA}$$

$$\text{Total loading} = 141 \text{ kW}$$

$$\text{Phase angle, } \phi = 23\cdot4°$$

∴ Resultant power factor, $\cos \phi = 0\cdot9245 \text{ lagging}$

5.8 Harmonic Currents

So far we have assumed for simplicity that a.c. invariably follows a pure sine wave and that at any instant of time the a.c. value is proportional to the mathematical sine of the angle. However, the wave form of the standard 50 Hz supply often deviates from the pure sine wave and is then known as a complex wave containing harmonics.

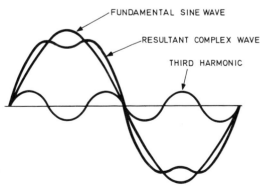

FUNDAMENTAL SINE WAVE

RESULTANT COMPLEX WAVE

THIRD HARMONIC

Fig. 5.15. A.C. wave with third harmonic.

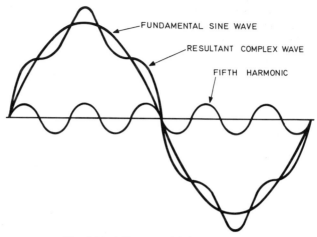

Fig. 5.16. A.C. wave with fifth harmonic.

The complex wave consists of a fundamental sine wave with harmonic sine waves; these latter have frequencies which are an exact multiple of the fundamental 50 Hz supply sine wave. In Fig. 5.15 it is seen that the complex wave is made up of a fundamental plus a third harmonic, while Fig. 5.16 shows the complex wave as derived from a fifth harmonic; the frequencies being 150 Hz and 250 Hz respectively.

The 15th Edition of the I.E.E. Wiring Regulations for the first time takes direct note of these harmonics. Regulation 331–1 requires that when planning an installation an assessment be made of the possible harmful effects from 'harmonic currents (such as fluorescent lighting loads and thyristor drives)'. Control gear from certain discharge lamps produces a third harmonic.

In a balanced 3-phase system with all phases following pure sine waves, the net current at the neutral is zero, but this is not the case with third harmonics present so that a current will now flow through the neutral. Under these conditions a reduction in the neutral conductor size is not permitted and must have the same cross-sectional area as the phase conductors.

EXERCISES

1. When a certain inductive coil is supplied at 240 V single-phase, 50 Hz, the current is 6·45 A. When the frequency is changed to 40 Hz at 240 V, the current taken is 7·48 A. Calculate the inductance and resistance of the coil.

2. A circuit consists of a resistance of 12 ohms, a capacitance of 320 micro-farads, and an inductance of 0·08 H, all in series. A supply of 240 V, 50 Hz is applied to the ends of the circuit. Calculate:
   (a) The current in the coil.
   (b) The potential differences across each element of the circuit.
   (c) The frequency at which the current would be unity power-factor. [C]

3. A choking coil of inductance 0·08 henry and resistance 12 ohm, is connected in parallel with a capacitor of 120 microfarads. The combination is connected to a supply at 240 V, 50 Hz. Determine the total current from the supply, and its power factor. Illustrate your answers with a phasor diagram.
                                                                              [C]

4. Three coils each with resistance, 45 ohms and with inductance 0·2 henry, are connected to a 415 V, 3-phase supply at 50 Hz, (a) in mesh, (b) in star.
   Calculate for each method of connexion (i) the current in each coil, and (ii) the total power in the circuit.   [C]

5. The installed load in a factory consists of 75 kW of 3-phase motors, 10 kW of tungsten lighting and 8 kW of 3-phase (non-inductive) heating.
   On the basis that (a) the supply is 240/415 V, 3-phase a.c., (b) the efficiency of the motors is 83%, their maximum demand is 60% of the installed load and their power factor is 0·8, while (c) the maximum demand of the lighting and heating is 100% of their installed loads, calculate:
   (a) the total load in kW and kVA,
   (b) the maximum line current, assuming that the 10 kW single phase lighting load is unbalanced to the extent of 2·8 kW on one phase, 3·2 kW on the second phase, and 4 kW on the third phase.   [C]

6. The following loads are supplied from a 415-V, 3-phase, 4-wire switch-board. The load, which is balanced on the three phases, comprises: (a) 160 kW at 0·7 power factor lagging, (b) 55 kW at 0·6 power factor leading and (c) 35 kW at unity power factor. Calculate the total load in kVA, the line current and the combined power factor.   [C]

7. A resistance of 24 Ω, a capacitance of 150 μF and an inductance of 0·16 H are connected in series with each other. A supply at 240 V, 50 Hz is applied to the ends of the combination. Calculate: (a) the current in the circuit, (b) the potential differences across each element of the circuit, (c) the frequency to which the supply would need to be changed so that the current would be at unity power-factor, and find the current at this frequency.   [C]

8. A 415-V, 3-phase, 4-wire system supplies power to three loads. The loads are: 20 kW at unity power factor between red and neutral; 30 kVA at 0·7 power-factor lagging between yellow and neutral; and 15 kW at unity power factor between blue and neutral. Calculate (a) the current in each line wire, (b) the current in the neutral conductor. Draw a phasor diagram illustrating your calculations.   [C]

Chapter 6

VOLTAGE DROP AND CURRENT RATING

6.1. General Considerations

From the fundamental consideration of circuit conditions, a voltage across a resistor causes current to flow through the resistor. When this condition occurs in main cables or long lengths of run it is often referred to as voltage drop, IR drop, or resistance drop. This voltage drop may also be seen as being due to the work that has to be expended in overcoming the opposition to the passage of current and must be deducted from the supply voltage in order to obtain the actual voltage at the load.

Since the cables are in effect a resistor—of low value—the entire arrangement may be seen as a simple series circuit. Figure 6.1 depicts the conditions where a voltage drop of 5 V occurs along cables fed from a 240 V circuit.

Let $\quad\quad\quad$ supply voltage $= V$

$$\text{Voltage across load} = V_2$$

$$\text{Voltage drop} = V_1$$

Then $\quad\quad\quad\quad\quad\quad V = V_1 + V_2$

i.e. $\quad\quad\quad\quad\quad\quad 240 = 5 + V_2$

$\therefore \quad\quad$ Voltage across load $V_2 = 235$ V

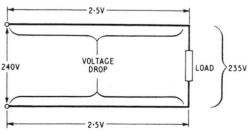

Fig. 6.1. Voltage drop seen as p.d.s in series with the load.

104

Where the load is reactive, phase differences must also be taken into account.

Example 6.1

Regulations specify maximum voltage drop limits for:
(a) motor circuits,
(b) lighting circuits,
(c) installations where equipment is installed to maintain constant voltage at main terminals.
Quote the figure for each of these.
Discuss the factors to be considered in calculating cable sizes for given loads. [C]

(a) The former permitted drop of $7\frac{1}{2}\%$ of the supply voltage, for motors, has now been replaced by the general figure of 2·5% of the declared or nominal voltage when the conductors are carrying full load current, but disregarding starting conditions. For satisfactory starting, the motor supply cables may have to be increased in size so as to permit a lower voltage drop than that given by the Regulations.

(b) The previous irrational figure of 2% + 1 V as the maximum permitted voltage drop from the consumer's terminals to any point in the lighting installation has now given way to a straight 2·5%. Thus, on a 240 V circuit, the voltage drop must not exceed 6 V.

I.E.E. Tables 9D1–9N1, subject to Rating Factors and Class of Protection as discussed below, show values of millivolt drop per ampere per metre of cable run (mV/A/m) when carrying the rated current.

Diversity $\left(\dfrac{\text{maximum load at one time}}{\text{total connected load}}\right)$ may also have to be taken into account, but this factor does not apply to conductors and switchgear of final circuits, except as specifically permitted by Regulation Appendix 4, Table 4B for cooking appliances.

(c) The Regulations do not now make any specific reference to voltage drop limits for installations design to maintain constant voltage at the main terminals.

The factors which must be taken into account when assessing cable sizes are discussed in the following sections.

6.2. Cable Selection

For safety purposes, statutory and I.E.E. Regulations stipulate that all cables must be sufficient in size for the current they have to carry.

The correct choice for any installation is dependent upon fundamental aspects of (a) environmental conditions and characteristics of protection, (b) current-carrying capacity and (c) voltage drop.

When current flows through a conductor, the resistance offered by the conductor produces heat. The increase in heat is proportional to the cable resistance which in turn depends upon the cross-sectional area of the cable. Since overheating damages the insulation, the conductor size must be of adequate size to prevent this from occurring.

6.3 Cable Protection

The current ratings of all acceptable types of cables are set out in the I.E.E. Tables as mentioned above and based on information supplied by the ERA (Electrical Research Association) Technology Ltd and the Electric Cable Makers' Confederation.

For determining the cable size with semi-enclosed (rewirable) fuses to BS 3036, the fuse rating must be divided by 0·725. Neglecting other considerations, the current-carrying capacity of 2·5 mm² p.v.c. twin and earth sheathed cable, clipped to the surface, as shown in Table 9D2 (col. 6) is 28 A. However, with protection by BS 3036 fuse, the cable must be sufficient in size to carry 30/0·725 A, i.e. 41·4 A. Hence to carry 28 A by this fuse, the cable must be uprated to 6 mm² (current-carrying capacity of 46 A).

This particular value of 0·725 does not apply to (i) cartridge type Class P or Class Q1 fuse-links, (ii) fuses with fuse-links complying with BS 1361, (iii) miniature and moulded circuit-breakers to BS 3871 and (iv) circuit-breakers set to operate at an overload not exceeding 1·5 times the designed load current of the circuit.

By now the reader will appreciate that the selection of an appropriate cable size from the 23 I.E.E. Tables of cable current ratings may be by no means a simple matter. In addition considerable modifications can occur to allow for *correction factors* determined by ambient temperature, grouping, type of sheath and disposition.

6.4. Ambient Temperature

If the temperature surrounding the cables is above or below 30°C allowance will have to be made in each case. The cables must carry a reduced current if the ambient or surrounding temperature is above 30°C; conversely, the cables may be up-rated for a reduced temperature. Ambient temperature includes the effect of artificial heat used in the building and even any local source of heat.

Example 6.2

From I.E.E. Table 9D1 *select cables of suitable current-carrying capacity for the following loads and conditions* (*p.v.c. cables to* BS 6004 *into screwed conduit*).

(*a*) 240 *V single-phase sub-mains of lighting load totalling* 10·5 *kW. Length of run* 10 *m. Average ambient temperature* 25°C, *diversity* 66%

(*b*) 400 *V balanced* 3-*phase power circuit. Load* 18·65 *kW, efficiency*, 80%, *power factor* 0·69. *Average temperature* 30°C. *Length of run* 100 *m.*

(*a*) Current taken by load	$= \dfrac{\text{Power}}{\text{Voltage}}$	
	$= \dfrac{10·5 \times 1000}{240}$	$= 43·75$ A
Permitted voltage drop	$= 2·5\%$ of 240 V	$= 6$ V
Allowing for diversity, maximum current through cables	$= 43·75 \times \dfrac{66}{100}$	$= 28·88$ A
(i) Protection by 30-A fuse to BS 88 Correction factor for 25°C	$= 1·06$	
∴ Required cable rating	$= \dfrac{30}{1·06}$	$= 28·3$ A
From Table 9D1, a 4 mm² conductor carries	32 A	
Testing for Voltage drop	$= \dfrac{\text{mV/A/m} \times I \times l}{1000}$	
	$= \dfrac{11 \times 28·8 \times 10}{1000}$	$= 3·168$ V

∴ Selected size is 4 mm²

(ii) Protection by 30-A fuse to BS 3036 requiring correction factor of 0·725 = 39 A

∴   Required cable rating $= \dfrac{30}{1\cdot06 \times 0\cdot725}$

From Table 9D1, a 6 mm² conductor carries 41 A

∴   Selected size is 6 mm²

(b)   Efficiency $= \dfrac{\text{Output}}{\text{Input}}$

$$= \dfrac{\text{kW} \times 1000}{\sqrt{3}VI \cos \phi}$$

$$\dfrac{80}{100} = \dfrac{18\cdot65 \times 1000}{\sqrt{3} \times 400 \times I \times 0\cdot69}$$

$$I = \dfrac{18\cdot65 \times 1000 \times 100}{\sqrt{3} \times 400 \times 80 \times 0\cdot69} = 48\cdot77 \text{ A}$$

Maximum voltage drop = 2·5% of 400 V   = 10 V

(i) Protection by 60 A circuit-breaker to BS 3871
From Table 9D1, column 3, 16 mm² cable carries 74 A as there are no correction factors

$$\text{Voltage drop} = \dfrac{\text{mV/A/m} \times I \times l}{1000}$$

$$= \dfrac{2\cdot3 \times 48\cdot77 \times 100}{1000} = 13\cdot17 \text{ V}$$

Too high

$$\text{Voltage drop for 25 mm}^2 = \dfrac{1\cdot7 \times 48\cdot77 \times 100}{1000} = 8\cdot29 \text{ V}$$

∴   Selected size is 25 mm²

Comment. This is one of the situations where the voltage drop becomes the main determining factor.

(ii) Protection by fuses to BS 3036

Cable must carry $\dfrac{60}{0\cdot725}$ A $= 82\cdot7$ A

Since 25 mm² cable carries 97 A,
selected size becomes 25 mm²

6.5. Grouping

With bunching, i.e. several cables run in one conduit, duct or trunking, an increase in temperature can arise, since each cable adds its quota of heat. A grouping correction factor also applies to p.v.c. twin and earth sheathed cables where laid closely together.

6.6. Sheathing

Sheathed multicore cables can be considered as a number of cables tightly bunched. There is thus a resemblance to grouping although no specific rating factors are provided. Instead reference to the I.E.E. Tables will show different ratings.

6.7. Disposition of Cables

Where it is clear that if they are placed vertically above each other, the cable above the bottom cable will experience some additional heat and this will increase with successive cables laid vertically above. A triangular or trefoil arrangement also reduces the rating, or the maximum current that the cables can safely carry. For each arrangement of the cables, including when clipped to one side or whether completely immersed in thermal lagging, one of the many correction factors must be applied.

Example 6.3

A 30 m run of twin and earth p.v.c. non-armoured copper cable is laid one side against thermal lagging and situated in an ambient temperature of 35°C. Determine the minimum size of cable to supply a 240-V 10-kW load. Protection given by (a) miniature circuit-breaker (m.c.b.) or (b) rewirable fuse.

$$I = \frac{P}{V}$$

$$= \frac{10\,000}{240} \qquad = 41\cdot67\,\text{A}$$

(*a*) 45 A m.c.b. is adequate for protection and from Table 9D2 correction factor for 35°C $\qquad = 0\cdot94$

Correction factor for cable in contact with lagging on one side (Reg. 522–6) $\qquad = 0\cdot75$

$\therefore$ Required cable current rating

$$= \frac{45}{0.94 \times 0.75} \qquad = 63.8 \text{ A}$$

From Table 9D2, 10 mm^2 cable carries 64 A

$$= \frac{\text{mV/A/m} \times I \times l}{1000}$$

$$= \frac{4.2 \times 41.67 \times 30}{1000} \qquad = 5.25 \text{ V}$$

$\therefore$ Selected cable size is 10 mm^2

(b) Required cable rating $= \dfrac{45}{0.94 \times 0.75 \times 0.725} \qquad = 88 \text{ A}$

From Table 9D2, 25 mm^2 cable carries 108 A

$\therefore$ Selected cable size is 25 mm^2

Comment. The example exhibits once more the considerable economic savings which can be gained by fitting an m.c.b. or correct cartridge fuse in place of the rewirable type.

6.8. Proximity of Cables to Steelwork

On occasion, allowance will have to be made for cables run near to steel, iron or ferroconcrete, due to the possibility of inductive effects.

6.9. Fundamental 3-phase Voltage-drop Calculations

These are all based on the basic formula $R = \rho(l/a)$ where ρ (rho) stands for resistivity. Resistivity is defined as the resistance between two opposite faces of a unit cube of the conductor material. Many voltage-drop problems involve the determination of resistance by this means and then multiplying by the current to obtain the IR drop. The weakness of this method, as against that adopted by use of the I.E.E. Tables, is that the Tables are much more realistic since they take into account the actual type of cable and conditions of service.

Example 6.4

A load of 300 *kW at* 0·78 *power-factor is to be supplied at* 415 *V,
3-phase, through a 3-core copper cable* 260 *m long. The cross-sectional
area of each cable core is* 400 *mm²*.
*Calculate the voltage drop in the cable. Ignore voltage drop due to
reactance. (Resistivity of copper may be taken as* 0·017 $\mu\Omega$-*m).*   [C]

Three-phase power, $P = \sqrt{3}VI \cos \phi$

By transposition $I = \dfrac{P}{\sqrt{3}V \cos \phi}$

$\qquad\qquad\qquad = \dfrac{300 \times 1000}{\sqrt{3} \times 415 \times 0\cdot78} \qquad = 535\cdot2 \text{ A}$

Resistance per core, $R = \rho\dfrac{l}{a} = \dfrac{0\cdot017 \times 260 \times 10^6}{10^6 \times 400} \qquad = 0\cdot01105 \ \Omega$

∴ Voltage drop in the cable

$\qquad\qquad\qquad = \sqrt{3} \times I \times R$

$\qquad\qquad\qquad = \sqrt{3} \times 535\cdot2 \times 0\cdot01105 \qquad = \underline{10\cdot24 \text{ V}}$

Comments. The important point to note is the voltage drop in a
3-core cable, when carrying the current due to a balanced 3-phase
load, is given by $\sqrt{3} \times IR$. The cable conductors are presumed to act
as a pure resistance without any reactive effects.

Example 6.5

Each core of a 3-core cable, 120 *m long, has a cross-sectional area
of* 10 *mm². The cable supplies power to a* 415-*V* 20 *kW 3-phase motor
working at full load with* 87 *per cent efficiency and at a power factor of*
0·72 *lagging. Calculate,*
(a) *the voltage required at the supply end,*
(b) *the power loss in the cable.*
(*The resistivity of copper may be taken as* 0·017 *microhm-metre*
($\mu\Omega$ m), *and the reactance of the cable may be neglected.*)
[C] as amended

(a)   Efficiency $= \dfrac{\text{output}}{\text{input}}$

i.e. $\dfrac{87}{100} = \dfrac{20 \times 1000 \text{ watts}}{\sqrt{3} \times 415 \ I \times \cos \phi \text{ watts}}$

$\therefore$ Line current $I = \dfrac{20 \times 1000 \times 100}{\sqrt{3} \times 415 \times 0 \cdot 72 \times 87}$ $= 44 \cdot 42$ A

Resistance of each core $R = \rho \dfrac{l}{a}$

$$= \dfrac{0 \cdot 017 \times 120 \times 10^6}{10 \times 10^6} \qquad = 0 \cdot 204 \ \Omega$$

(note division by 10^6 to change mm² to m²).

Power loss per core $= I^2 R$

$$= 44 \cdot 42^2 \times 0 \cdot 204 \qquad = 402 \cdot 5 \ \text{W}$$

Power loss in 3-core cable

$$P_c = 402 \cdot 5 \times 3 \qquad = 1 \cdot 208 \ \text{kW}$$

motor power input $= \dfrac{\text{output}}{\text{efficiency}}$

$$= \dfrac{20 \ \text{kW}}{0 \cdot 87} \qquad = 23 \ \text{kW}$$

The problem may now be set out in the form of a phasor diagram (Fig. 6.2).

Now $\qquad\qquad \cos \phi = 0 \cdot 72$

$\therefore \qquad\qquad\qquad \phi = 43° \ 56'$

also $\qquad\qquad \dfrac{\text{kVAr}}{\text{kW}} = \tan \phi$

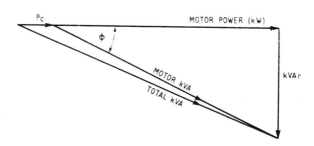

Fig. 6.2. 3-phase voltage drop power diagram. P_c = power loss in 3-core cable.

$$\therefore \quad \text{kVAr} = \text{kW} \tan \phi$$

$$= 23 \times 0.9634 \qquad = 22.16 \text{ kVAr}$$

Total power from supply = power lost in 3-core cable + motor power

$$= 1.208 + 23 \qquad = 24.208 \text{ kW}$$

$$\text{Total kVA} = \sqrt{\{(\text{total power})^2 + (\text{kVAr})^2\}}$$

$$= \sqrt{(24.208^2 + 22.16^2)} \qquad = 32.82 \text{ kVA}$$

$$\text{kVA} = \frac{\sqrt{3}VI}{1000}$$

$$\therefore \quad \text{Supply voltage } V = \frac{1000 \text{ kVA}}{\sqrt{3}I}$$

$$= \frac{1000 \times 32.82}{1.732 \times 44.42} \qquad = 427 \text{ V}$$

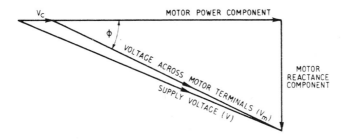

Fig. 6.3. 3-phase voltage phasor diagram. V_c = voltage drop in 3-core cable.

ALTERNATIVE SOLUTION. This makes use of the voltage phasor diagram. The motor load voltage (V_m) can be split into two right-angle components—in-phase or power and quadrature or reactive. Referring to the phasor diagram (Fig. 6.3), and working in 3-phase values

Voltage drop in cable $\quad V_c = \sqrt{3}IR$

$$= \sqrt{3} \times 44.42 \times 0.204 \qquad = 15.7 \text{ V}$$

Power component of load $= V_m \cos \phi$

$$= 415 \times 0.72 \qquad\qquad = 298.8 \text{ V}$$

This is *in phase with the cable voltage drop.*

Total power component $= 15.7 + 298.8 \qquad = 314.5 \text{ V}$

Reactive component $= V_m \sin \phi$

$$= 415 \times 0.694 \qquad\qquad = 288 \text{ V}$$

Supply voltage $= \sqrt{\text{(total power component)}^2}$
$+ \text{(reactive component)}^2$

$$= \sqrt{(314.5^2 + 288^2)} \qquad = 427 \text{ V}$$

Comment. The voltage phasor diagram is derived from the power diagram as the value represented by each side is divided by the current it is exactly the same shape, but could be drawn to a different scale. The voltage diagram has the advantage of showing clearly that the supply voltage is the *phasor sum* of the cable voltage drop and the voltage at the motor terminals. Another way at looking at the problem is to regard the cable as a resistor and the induction motor as a coil, which when combined produce the equivalent circuit of Fig. 6.4.

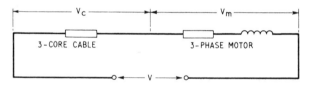

Fig. 6.4. 3-phase equivalent circuit.

Example 6.6

The main switchboard of a works supplies a workshop 240 m distant, with 3-phase power for a balanced load of 80 kW at 0.7 power factor lagging. The voltage at the load to be maintained at 415 V, 3-phase.

The volt drop in the cable must not be more than $7\frac{1}{2}\%$ of the voltage at the main switchboard. Calculate the minimum size of cable needed for this purpose. If the size thus found is not included in the range of standard sizes given below choose the most suitable cable size and calculate the actual voltage drop in the cable.

(The resistivity of copper may be taken as 0.017 microhm-metre.)

Standard sizes of cables (nominal cross-sectional area in mm²): 10, 16, 25, 35, 50, 70.   [C] as amended

$P = \sqrt{3}\, V_m I \cos\phi$ where V_m = voltage at motor terminals

$$\text{Line-current } I = \frac{P}{\sqrt{3}\, V_m \cos\phi} = \frac{80\,000}{\sqrt{3} \times 415 \times 0.7} = 159.2\text{ A}$$

Let the voltage at the switchboard be V_s.

Then, maximum voltage drop along cables

$$V_c = 7\tfrac{1}{2}\% \text{ of } V_s = \frac{7\frac{1}{2}V_s}{100} = 0.075V_s$$

As $V_m = 415 = V_s - V_c = V_s - 0.075V_s = 0.925V_s$

∴ $V_s = \dfrac{415}{0.925}$ $= 448.6$ V

Voltage drop along 3 cable cores

$= 7\tfrac{1}{2}\%$ of 448.6 $= 33.65$ V

But 33.65 V $= \sqrt{3} \times$ volt drop/core

∴   Volt drop/core $= \dfrac{33.65}{\sqrt{3}}$ $= 19.43$ V

∴ Resistance/core, $R = \dfrac{\text{volt drop/core}}{\text{line current}} = \dfrac{19.43}{159.2}$ $= 0.122\ \Omega$

Now $R = \rho\,\dfrac{l}{a}$

∴ $a = \rho\,\dfrac{l}{R} = \dfrac{0.017 \times 240 \times 10^6}{10^6 \times 0.122}$ $= 33.44\text{ mm}^2$

(Multiplier 10^6 to convert to mm².)

Selected cable size is 50 mm²

Adopting this size, voltage drop

$= \sqrt{3}\,IR$

$= \dfrac{\sqrt{3} \times 159.2 \times 0.017 \times 240 \times 10^6}{10^6 \times 50}$ $= \underline{22.5\text{ V}}$

Comments. Strictly speaking, the phase difference between the voltage at the terminals and the voltage drop along the cables should be taken into account. However this may be neglected as due allowance has been made in selecting the nearest size cable.

The 3-phase volt-drop calculations could also be calculated by use of the I.E.E. Tables, but in this case more information would be required in order to select the appropriate type of cable for conditions of service.

EXERCISES

1. (a) In assessing the size of conductors, explain what the I.E.E. Regulations stipulate with respect to:

(i) maximum permissible voltage drop at any point on installations (other than motor circuits) and, alternatively, on installations without regulators that maintain an approximate constant voltage at the consumer's main terminals,

(ii) the maximum permissible voltage drop at motor terminals,

(iii) the current carrying capacity of the conductors.

(b) Explain the effect of ambient air temperature on the current ratings of vulcanised rubber and p.v.c. insulated cables.   [C]

2. A 415 V, 40 kW, 3-phase motor has an efficiency at full load of 85% and a power factor of 0·8 lagging. The mineral insulated cable feeding the motor is 100 m long and each conductor has a cross-sectional area of 10 mm².

(a) Calculate the voltage drop in the cable (the resistivity of copper may be taken as 0·017 $\mu\Omega$-m).

(b) What are the limits of the voltage drop allowed by the I.E.E. Regulations for this circuit?   [C] as amended

3. A 100 kW, 3-phase motor working at full load at a power factor of 0·75 with an efficiency of 87%, is supplied at 415 V from a works main switchboard by means of a 3-core cable 150 m long. Each core of the cable has a cross-sectional area of 50 mm². Calculate (a) the voltage at the switchboard, (b) the power losses in the cable. The resistivity of copper may be taken as 0·017 $\mu\Omega$ m.   [C] as amended

4. (a) Calculate the effect on electric heaters of a 10% voltage drop.

(b) Explain whether the type of cable sheathing or ambient temperature affects voltage drop.

(c) In what way does the surrounding temperature affect the current-carrying capacity of cables?

5. It is required to transmit 50 kW at 400 V between lines at a point 400 m distant. Assuming the same size of cable to be used in each case, which would be the more economical system to use, d.c. or 3-phase a.c.?

6. A 3-phase balanced load of 180 kW, working at 415 V and at a power factor of 0·75 lagging, is supplied from a works switchboard by means of a 3-core cable. The resistance and inductive reactance are respectively 0·015 Ω and 0·005 Ω.

(a) Calculate the voltage at the switchboard.

(b) Draw a typical phasor diagram (not to scale) representing the above.

7. Four 500 W motors at an efficiency of 85% each, twelve 2 kW radiators and 100—80 W lamps are fed by PVC armoured cable, clipped to the surface, from a point 100 m away. If the mains are subject to a temperature of 40°C, neglect the effect of power factor and calculate the cable size if the supply voltage is 240 V. Allow a maximum voltage drop of 5 V.

8. (a) What are the requirements of the I.E.E. Regulations regarding voltage drop in an installation?

(b) It is desired to take a 240-V single phase a.c. supply to a 15-kW load, using a 16 mm² two-core m.i.c.c. cable. What is the maximum length of run of allowed protection by fuse to BS 1361, Pt. 2?

(c) If the cable passes through an area with an ambient temperature of 50°C, what is the maximum current which can be carried under these conditions?   [C & G]

Chapter 7

DISTRIBUTION

7.1. Generation

Electricity supply for the overwhelming majority of installations in Britain comes, in the first instance, from some 300 highly-efficient power stations. Most of these are still coal-fired, for the purpose of raising the steam which in turn drives the turbo-generators. Crude oil is used for diesel engines, in certain cases, where plentiful supplies of this fuel are readily available. In the mountainous regions of North Wales and Scotland water power by means of dams produces hydro-electricity. The future of electricity generation appears to lie in nuclear power and already a number of such stations are in active commission. Uranium rods are employed as the 'fuel' for these nuclear reactors. Under controlled conditions the rods produce intense heat to raise steam and drive conventional generators. There is also work going on by scientists for the production of electrical energy by simpler means and so to avoid the expensive and complicated generating plant and even to eliminate the need for transmission. They talk about a 'black box' that will supply electricity direct from the atom but whether such possibilities will be realised appear to lie very much ahead in time.

7.2. The Grid

For the present at any rate, it is a fundamental fact that the higher the voltage, the cheaper becomes the supply. Since $P = VI \cos \phi$ watts, it follows that an increased voltage makes for a reduced current for a given amount of power and power factor. Under such circumstances we can state that the current is inversely proportional to the voltage. *A lower current obviously results in reduced cable and switchgear sizes, $I^2 R$ losses and voltage drop.* This is the reason for the trend in adopting higher and higher voltages.

For practical reasons the limiting voltage available from present-day alternators (as the a.c. generators are called) is about 20 000 V. Transformers are thus necessary to step up the electricity up to the highest possible voltage consistent with safety and transmission technique. The general grid voltage of 132 kV has been augmented by a stiffening pressure of 400 kV forming a Super-Grid running from north to south of England. Even this voltage has been increased by lines fed at 750 kV and plans are afoot for even higher voltages.

Fig. 7.1. Fourteen Area Electricity Boards.

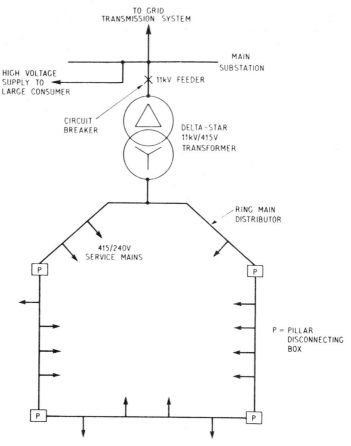

Fig. 7.2. Typical distribution system.

This may appear to be a far step from the voltages with which the electrician has to deal. Nevertheless certain high voltages may come within his sphere as the larger factories are often fed at 11 000 volts and in any case he should consider the part that l.v. (low voltage) and m.v. (medium voltage) plays in the general scheme.

From the main Grid network at selected transformer stations (usually outdoor), the voltage is reduced to 11 kV at which pressure current is purchased by the 14 Electricity Boards (Fig. 7.1) from the Central Electricity Board which is responsible for the Grid and the entire generation in Great Britain. There is a secondary network, formerly at 6·6 kV but now mostly 11 kV, which serves local sub-stations. Most consumers are fed from these substations at step-down standard voltages of 415/240 V, 50 Hz. As already stated the larger industrial plants may take their supply directly at 11 kV and provide their own private substation. A typical part of the national transmission and distribution system is shown in Fig. 7.2.

The underground distributor supplies the service mains, which are 4-wire, for urban, i.e. town or city, consumers. The same principle applies to rural distribution, but here the cables would probably be run overhead on poles.

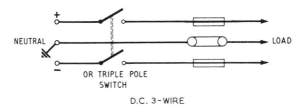

OR TRIPLE POLE
SWITCH

D.C. 3-WIRE

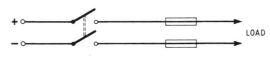

D.C. 2-WIRE (NEITHER POLE EARTHED)

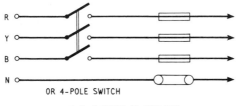

OR 4-POLE SWITCH

A.C. 4-WIRE (3-PHASE)

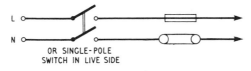

OR SINGLE-POLE
SWITCH IN LIVE SIDE

A.C. 2-WIRE (SINGLE PHASE)

Fig. 7.3. Supply controls.

7.3. Supply Voltages and Control

The relevant I.E.E. definitions state the range of voltages for standard use:

EXTRA LOW (e.l.v.). 0–50 V a.c. or 0–120 V d.c. between conductors or to earth.

LOW (l.v.). Above e.l.v. to 1000 V a.c. (1500 V d.c.) between conductors or 600 V a.c. (900 V d.c.) between conductors and earth.

EXTRA HIGH PRESSURE. This term is recognised by the Electricity Factory Acts for voltages above 3000 V.

Installation supply voltages are all subject to the variations as permitted by the Electricity Supply Regulations, 1937 (E.S.R.) which is $\pm 6\%$. Thus these statutory limits for 240 V are from 225·6 V to 254·4 V. It is not often realised that the permitted frequency variation is $\pm 1\%$. The controls for the principal l.v. systems are shown in Fig. 7.3.

7.4. D.C. 3-wire Distribution

Whilst little used for mains supply now, the calculations provide useful exercises towards understanding the basic differences between potential and p.d. The general scheme is illustrated in Fig. 7.4.

Larger apparatus such as 480 V motors are fed directly from the outers, some of the l.v. loads being taken directly off the positive outer and the mid-wire while others are fed from the mid-wire to the negative outer. The neutral mid-wire is marked $\pm$ (plus or minus) as it is negative with respect to the top wire in the diagram and positive with respect to the bottom one.

BALANCING. The loads are aimed to be balanced, the mid-wire carrying the difference or out-of-balance current between the outers. For this reason its cross-sectional area is often reduced to about half. Any unbalance would produce uneven voltages across the loads. The purpose of the balancer is to redress this unbalance. Referring to Fig. 7.4, X and Y are the armatures of two d.c. machines connected in series across the outers, the fields being cross-connected. The set is started up as two machines across the outer wires. Out-of-balance loads may make each machine motor or generate automatically in order to make up for voltage drop or rise.

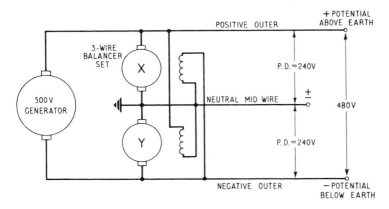

Fig. 7.4. D.C. 3-wire distribution.

If the positive outer is more heavily loaded it will produce a drop in voltage with respect to the neutral and a corresponding rise in voltage in the other side. Thus machine Y will have a higher voltage across its armature than machine X; however due to the cross-connexion, its field will suffer a drop. This will result in a voltage rise due to the reduced excitation and cause armature X to generate an increased voltage. A corresponding action occurs with increased negative load.

7.5. 3-wire D.C. Calculations

As already mentioned the solution of 3-wire d.c. problems requires an appreciation of potential and potential difference. Assuming an earthed system, point A (Fig. 7.5) has a potential of 240 V with respect to earth whilst point D is at zero potential. Similarly points B and C are assumed to be at potentials of 175 V and 125 V respectively, but the potential difference between B and C = 175 − 125 = 50 V.

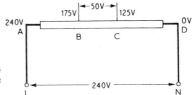

Fig. 7.5. Points B and C are at potentials of 175 V and 125 V respectively but the p.d. between B and C is 50 V.

Example 7.1

A 3-wire d.c. system supplies power to two adjacent workshops by means of a 3-core aluminium cable, 300 m long. The voltage at the main switchboard is maintained at 240/0/240 V.

Workshop A takes 200 A from positive and neutral, and workshop B takes 250 A from neutral and negative. The cross-sectional area of each outer core of the cable is 320 mm², and the cross-sectional area of the inner core is 200 mm².

Show on a suitable diagram the values and directions of the currents in the cable, and calculate the voltages at A and B respectively.

Describe in general terms what would happen if the neutral core became disconnected.

The resistivity of aluminium may be taken as 26·7 μΩ mm.   [C]

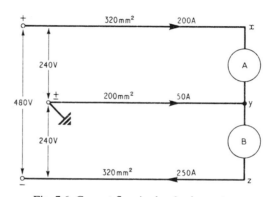

Fig. 7.6. Current flow in d.c. 3-wire system.

From Kirchhoff's current law, the out-of-balance current in the neutral wire is 50 A in a direction outwards from the supply (Fig. 7.6).

$$\text{Resistance of each outer} = \rho\,\frac{l}{a} = \frac{26\cdot7 \times 300 \times 10^3}{10^6 \times 320} \qquad = 0\cdot025\ \Omega$$

$$\text{Resistance of neutral} \qquad = 0\cdot025 \times \frac{320}{200} \qquad = 0\cdot04\ \Omega$$

Volt drop along +ve outer = 200 A × 0·025   = 5 V

Volt drop along neutral   = 50 × 0·04   = 2 V

Volt drop along −ve outer = 250 × 0·025   = 6·25 V

Potential of x above earth = 240 − 5   = 235 V

Potential of y below earth	$= 0-2$	$= -2$ V
Voltage at A	$=$ p.d. between x and y	
	$= 235-(-2)$	$= 237$ V
Potential at z	$= -240+6\cdot25$	$= -233\cdot75$ V

(N.B. The positive 6·25 V represents a voltage rise)

Voltage at B	$=$ p.d. between y and z	
	$= -2-(-233\cdot75)$	$= 231\cdot75$ V

In the event of a break in the neutral, the load would be placed in series with the 480 V supply. Thus, the voltage across each load would be in direct proportion to the resistance of each load.

7.6 Distributor Fed at Both Ends

A 2-wire distribution cable, ABCDE 150 m long and of uniform cross-sectional area, is supplied at both ends A and E at 250 V. Consumers' loads are connected as follows: 30 A at B, 40 m from A; 45 A at C, 100 m from A; 75 A at D, 120 m from A. The total resistance of the cable, lead, and return is 0·12 Ω. Find:

(a) the value of the currents entering the cable at A and at E;
(b) the potential difference at point C;
(c) the power loss in the cable.   [C]

(a) By setting out the loads and distances (Fig. 7.7(a)) the resistance of each cable section can be obtained by being considered as a proportion of the total resistance:

$$R_{AB} = \frac{40}{150} \times 0\cdot12 = 0\cdot032\Omega$$

$$R_{BC} = \frac{60}{150} \times 0\cdot12 = 0\cdot048\Omega$$

$$R_{CD} = \frac{20}{150} \times 0\cdot12 = 0\cdot016\Omega$$

$$R_{DE} = \frac{30}{150} \times 0\cdot12 = 0\cdot024\Omega$$

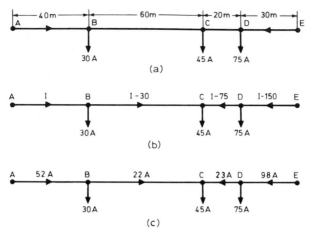

Fig. 7.7. (a) Distributor fed at both ends. (b) Current values in terms of I (amperes). (c) Section current values and directions.

Let the current through AB be I amperes, then the current values in each section can be stated in terms of I (Fig. 7.7b).

From Kirchoff's second law (see 4.6.) the sum of the potential differences are equal to zero i.e.

$$0.032I + 0.048 (I - 30) + 0.016 (I - 75) + 0.024 (I - 150) = 0$$

so that

$$0.12I - 6.24 \qquad = 0$$

and

$$I = \frac{6.24}{0.12} \qquad = 52 \text{ A}$$

All current values can now be inserted in the line diagram, noting that a negative current value signifies a reversal of current direction (Fig. 7.7c).

(b) p.d. at C = p.d. at A − section AB voltage drop − section BC voltage drop

$$= 250 - (52 \times 0.032) - (22 \times 0.048)$$
$$= 250 - 1.66 - 1.06$$
$$= 247.26 \text{ V}$$

(c) Power loss in sections $(I^2R) = 52^2 \times 0.032$ $= 86.53 \text{ W}$
$$= 22^2 \times 0.048 \qquad = 23.32 \text{ W}$$
$$= 23^2 \times 0.016 \qquad = 8.46 \text{ W}$$
$$= 98^2 \times 0.024 \qquad = 230.49 \text{ W}$$
Total power loss $= 349 \text{ W}$

7.7 Ring Main Distribution

Similar in many respects to the domestic 30-A ring circuit, the method has many advantages over the radial system. Fullest use is made of conductor cross-sectional area due to diversity applied to the loads and normally results in significant cable economies.

Example 7.3

A factory 11 kV ring main distribution system has four low-voltage 500 kVA load centres and is fed at one point.

(a) Draw a single-line diagram of the ring commencing with the supply and finishing at each of the four transformers, showing all necessary protection.

(b) Draw a single-line diagram of a typical load centre showing the high voltage and low-voltage switchgear and transformer, assuming six outgoing circuits.

(c) Discuss the factors affecting the choice of switches or circuit-breakers for ring-main isolation.

[Inst. Tech. Pt. 2]

(a) Fig. 7.8(a) illustrates the appropriate line diagram for the ring main with an individual switchfuse or circuit-breaker for each transformer. For purposes of isolation additional switchfuses/circuit-breakers could be positioned in each section of the ring.

(b) Layout shown in Fig. 7.8(b).

(c) The fitting of switchfuses in place of circuit-breakers would show an initial saving in costs, although with present-day prices this differential is narrowing. However, heavy maintenance charges may favour the switchfuse.

Circuit-breakers allow for the rapid replacement of supply after clearance of faults. They also permit greater flexibility so that, for example, relays could be fitted which would automatically isolate faulty section of the ring or parts of the outgoing circuits.

7.8. Private Substations

For large total loads, commercial or industrial, transmission may be brought in at 11 kV and a substation will be required on the premises. The actual intake position must be decided in conjunction with the local Area Board. The most economic position will be as near to the centre of the load as possible. This would minimise

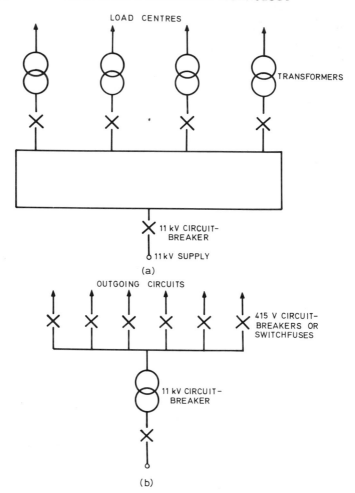

Fig. 7.8. (a) Ring main with transformer-fed load centres. (b) Supply to outgoing circuits.

undue lengths of heavy cable runs and reduce voltage drop and power losses.

A typical line diagram of the main electrical connexions and substation layout may be seen in Fig. 7.9. The floor area is about 20 m² and height 2·75 m. Since lifting tackle may be used to place the various items of equipment prior to grouting in position, the roof may have to be strengthened for this purpose. The main doors must be of sufficient size to take the transformer, to avoid the wall of the

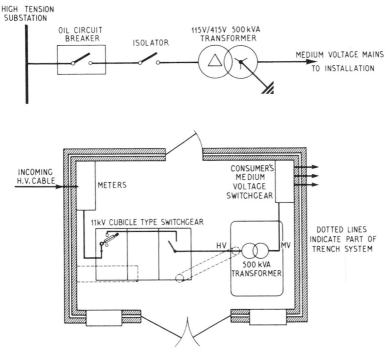

Fig. 7.9. Line diagram of main connexions and typical layout of private substation.

substation having to be cut away later to allow for its entry. Both doors should open outwards to give access to the road.

Ventilation of the substation is essential to prevent condensation and some form of artificial heating may be necessary to prevent the collection of moisture during winter. Whilst the maximum amount of natural lighting is desirable, some form of secondary lighting is an obvious advantage.

FIRE RISK. This hazard is mainly caused through the possibility of igniting the oil used in the transformer and circuit breakers. Oil can be set alight by open faults in the windings. Explosions can also occur due to heavy transformer short-circuits. Also oil may be vaporised by arcing or loose connexions, the gas thus formed mixing with the air to produce an explosive atmosphere.

Any spilled oil should be suitably contained. There is a real danger of burning oil escaping from the oil tank and starting an oil fire which could be almost impossible to control. Thus, means must be adopted to prevent an unrestricted flow. For this purpose, it is common to bed ducts with small granite chippings. In some cases, the ducts lead to a common larger duct outside the building to allow for effective draining away of surplus oil.

Here I.E.E. Regulation 422–5 is worth quoting. 'Wherever electrical equipment in a single location contains flammable delectric liquid in excess of 25 litres, precautions shall be taken to prevent burning liquid and the products of combustion of the liquid (flame, smoke, toxic gases) spreading to other parts of the building.'

Note. Examples of precautions to be taken are given in Chapter 11, section 11.14.

A typical arrangement is to lay the cables in trenches some 0·76 m deep, to allow for wide sweeps which should have a minimum radius of 0·76 m, and the trenches should be covered with steel chequer-plates.

CO_2 (carbon dioxide) is considered the most effective means for combating any outbreak of fire. The gas should be incorporated as part of an automatic fire-fighting system.

For centralised power control, a new trend has developed in the form of 'packaged' substations. This incorporates a *dry-type trans-former* which avoids the fire and explosive hazards of its oil-filled counterpart. It also includes air-break circuit-breakers up to 1600 A, requiring only a simple mechanical erection.

7.9 Safety to Personnel at High Voltage

Both the statutory Electricity (Factory Acts) Special Regulations, 1908–44 and the Electricity Supply Regulations, 1937 should be closely studied. Regulation 10(*a*) of the latter is appropriate to this subject and is worth quoting in full.

All doors or covers shall be so secured that they cannot be opened except by means of a key or special appliance. The enclosed conductors and apparatus shall be so constructed, protected and arranged that when the door or cover giving access to an operating or switch panel is opened, it shall not be possible for the person opening the door or cover to come in accidental contact with metal electrically charged at high voltage. Unless the conditions of supply are such that the whole of the enclosed conductors and apparatus may be made dead at the same time for the purpose of cleaning or other work thereon, the conductors and apparatus shall be so arranged that they made dead in sections, and the sections shall be so separated by divisions or screens from all adjacent live metal that work on any section made dead may be carried on by an authorised person without danger. Every fusible cut-out shall either be capable of being made dead by a switch or shall be so con-

structed and placed that it can be handled without danger by an authorised person for the purpose of renewal.

Electrical shock injury or burns may be as a result of working with the supply connected or with the accidental switching on under these conditions. It is thus vital to adopt a rigid procedure to prevent such occurrences. Access to the substation must only be possible to authorised persons. The Factory Acts, as quoted above, state that such persons shall periodically examine the insulating stands, screens, rubber boots and gloves which must be provided. Instruc-

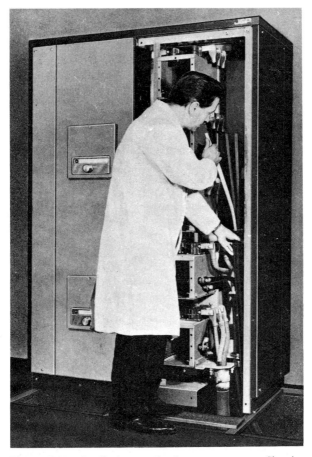

Fig. 7.10(*a*). Small factory intake arrangement. Showing medium-voltage fuse-switchgear with universal access. Cabling is normally carried out from the front and the switchboard can be positioned against a wall. (Johnson & Phillips Ltd.)

tions, such as are obtainable from the *Electrical Times* or *Electrical Review*, for the treatment of persons suffering from electric shock must be fixed in a clearly visible position.

7.10. Cubicle Switchgear

The metalclad switchgear in substations and other large intake positions was formerly mounted on stout angle-iron framing and normally made up on site. This has almost wholly been replaced by switchgear enclosed in a cubicle which may be of the truck type to facilitate maintenance.

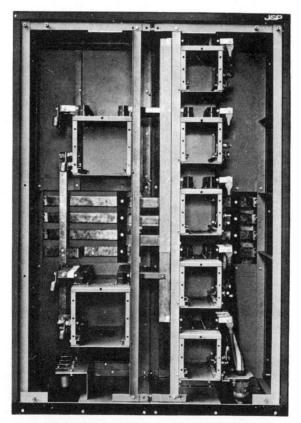

Fig. 7.10(*b*). Front view of m.v. fixed-cubicle intake arrangement of (*a*), showing mounting of units and cable access with doors open. (Johnson & Phillips Ltd.).

The circuit-breaker and current and voltage transformers are mounted on a truck which can be pushed into the cubicle. A partition with automatic shutters prevents access to busbars when the truck is withdrawn. A mechanical interlock is fitted to prevent the truck from being withdrawn until the supply is ruptured. Two views of a medium-voltage fixed-cubicle type are shown in Fig. 7.10(a–b). The clean finish is apparent and the unit can accommodate five 100 A or two 800 A fuse-switches. There are also facilities for distribution boards and metering. As typical of modern switchgear technique, the operation is by manual spring-operated mechanism which removes the danger of hesitant action on the part of the operator. The springs are charged manually but the speed of opening or closing is completely independent of the operator.

7.11. Distribution of Low Voltage Supplies

DIVERSITY AND MAXIMUM DEMAND. When calculating the necessary currents for switchgear, mains or submains in any installation, simply to add all connected loads would usually be impracticable and too costly, the reason being that it is unlikely that the loads will all be used together. The loads are thus diversified which has led to the term diversity, or diversity factor (D.F.) when expressed as a percentage of the connected load.

D.F. may be applied to installations in order to estimate the maximum demand (M.D.) at any time. Here a warning must be given that the actual assessment of diversity calls for care and experience. A general guide is given in Appendix 4 of the I.E.E. Wiring Regulations. In notes on the application of diversity it states that the amounts 'may be increased or decreased as decided by the engineer responsible for the design of the installation concerned.' Diversity is not applicable to final sub-circuits, the exception being the 30-A ring circuit. Too high a D.F. will result in larger fusegear, busbars, etc. than necessary; too low a D.F. must result in insufficient provision for future supplies.

Diversity may also occur in time by referring to when peak periods do not coincide, or to groups of loads such as blocks of flats.

RISING MAINS. Part of a proprietary scheme suitable for new flats and other commercial buildings may be seen in Fig. 7.11. The unit which is continued throughout the height of the building, is fitted into recesses left by the builders.

Generally, conductors employed for main distribution in multistorey buildings consist of solid copper or aluminium round or

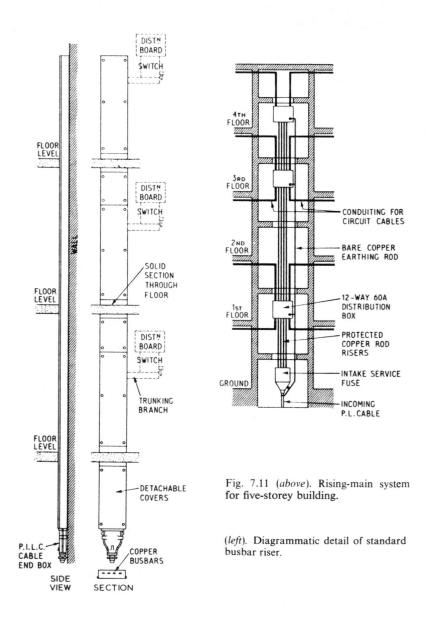

Fig. 7.11 (*above*). Rising-main system for five-storey building.

(*left*). Diagrammatic detail of standard busbar riser.

rectangular rods suitably fixed to insulators and enclosed in a rust-less sheet steel trunking.

Light insulation is sometimes employed mainly for the purpose of closer spacing in order to reduce the overall dimensions of the casing. One maker has enabled a further reduction to be made by the use of plastics insulation between the conductor bars.

SUPPLY. The feed point is usually at the bottom but, where the building is exceptionally high or the load is unduly great, the bars may be cut across in half and both halves fed separately. This achieves a 100 % increase in current-carrying capacity. Skyscraper buildings may necessitate substations at the mid-point of the building, to reduce voltage drop.

An alternative scheme for increasing the current rating is to bring up two sets of risers which are then connected to form a giant ring main circuit.

Covering the bare conductors with black paint increases their current rating by 20 %. The use of split parallel bars is another way to increase the current-carrying capacity, as this increases the surface area and therefore allows greater heat dispersal.

There is a wide range of busbar current sizes, 200 A to 600 A being the range available from a number of suppliers. The actual load is found by compiling a complete schedule of the loads. Then the total connected load is multiplied by a diversity factor. The nearest busbar rating is then chosen, allowing for voltage drop, and, if necessary, temperature rise. BS 159 stipulates that this rise shall not be more than 50°C with an average surrounding air temperature of 35°C. Four bars are usually fitted for a triple pole and neutral arrangement, but two or three bars are sometimes installed. Standard lengths are 2 m and 4 m, although intermediate lengths are obtainable.

For installation purposes, the building contractor must leave appropriate chases and it is advisable to install the steel trunking as soon as the roof is put on or the site rendered waterproof. By being 'built in', solid support is then largely provided for the not-inconsiderable weight of the rising main.

7.12. Requirements for Bare Conductors in Trunking

The conductors must be inaccessible to unauthorised persons or totally enclosed in earthed metal or fixed in a chase, channel, trunking or shaft specially provided for the purpose.

The conductors and insulators are required to be of adequate strength to withstand the electro-mechanical forces which may be set up by the prospective short-circuit current (see Fig. 1.15). They must be free to expand without damage due to temperature changes and at each straining position, suitable straining gear with double insulation must be provided.

To prevent the spread of fire and the top of the trunking attaining excessive temperatures, suitable fire-resisting barriers must be provided.

Example 7.4

A 6-storey building of 24 flats, each equipped for 'all-electric' working, is supplied at 230/400 V 3-phase 4-wire. The installed load in each flat is 9·5 kW. The installed load for landlord's lighting is 2·4 kW, and is controlled by a clock switch to operate between sunset and sunrise. Two passenger lifts are each powered by a 6 kW motor. Determine:

    (a) the total installed load,
    (b) the M.D. on the service cable, assuming a diversity factor of 65% on the 24 flats,
    (c) the capacity of main switch fuse in each flat,
    (d) the capacity of the main isolator at intake position.   [C]

(a) Total installed load

Flats $24 \times 9\cdot5 = 228 \text{ kW}$

Landlord's lighting $= \quad 2\cdot4 \text{ kW}$

Lift motors

$$\text{Assuming } 90\% \text{ efficiency} = \frac{2 \times 6000 \times 100}{90 \times 1000} = \underline{13\cdot33 \text{ kW}}$$

$$\text{Total} = \underline{\underline{243\cdot73 \text{ kW}}}$$

(b) Maximum demand

Flats $228 \times 65/100 = 148\cdot2 \text{ kW}$

Landlord's lighting $= \quad 2\cdot4 \text{ kW}$

Lift motors $= \quad \underline{13\cdot33 \text{ kW}}$

$$\text{Total} = \underline{\underline{163\cdot93 \text{ kW}}}$$

Note: The whole of the landlord's lighting and lift loads will be switched on so that there is no diversity allowance.

(*c*) Assuming D.F. of $66\frac{2}{3}$ %

$$\text{Current} = \frac{9 \cdot 5 \times 66 \cdot 67 \times 1000}{230 \times 100} = 27 \cdot 54 \text{ A}$$

∴   30 A main switch fuse suitable

(*d*) The flats are assumed to be spread over 3 phases

∴ $$\text{Power/phase} = \frac{148 \cdot 2}{3} = 49 \cdot 4 \text{ kW}$$

∴ $$\text{Current/phase} = \frac{49 \cdot 4 \times 1000}{230} = 214 \cdot 8 \text{ A}$$

$$\text{Current for landlord's lighting} = \frac{2 \cdot 4 \times 1000}{230} = 10 \cdot 43 \text{ A}$$

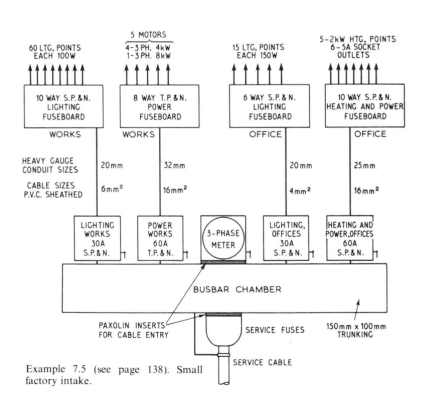

Example 7.5 (see page 138). Small factory intake.

Lift motors

$$\text{Efficiency} = \frac{\text{output}}{\text{input}} = \frac{90}{100} = \frac{12\,000}{400 \times I \times 0\cdot9}$$

(assuming power factor of 0·9)

Hence, lift current,

$$I = \frac{12\,000 \times 100}{\sqrt{3} \times 400 \times 90 \times 0\cdot9} \qquad = 21\cdot4 \text{ A}$$

$$\text{Total} \qquad 246\cdot6 \text{ A}$$

∴   Size of main isolator 300 A triple pole and neutral

Example 7.5

Make a line drawing of the circuit arrangements and control gear for an installation in a small factory comprising:
    (i) *sixty lighting points in the works at 100 watts each.*
    (ii) *one motor in the works at 8 kW (p.f. 0·8),*
    (iii) *four motors in the works at 4 kW (p.f. 0·7),*
    (iv) *fifteen lighting points in the offices at 150 watts each,*
    (v) *five heating points in the offices, 2 kW load to each outlet,*
    (vi) *six socket outlets in the offices for 0·5 kW motors.*
Supply: 3 phase 240 V/415 V. Metering: two-part tariff.

EXERCISES

1. What are the requirements for installing sheathed or armoured cables in rising ducts with particular reference to the following: (a) air temperatures, (b) spread of fire, (c) fixings, (d) damage to insulation.   [C]

2. A factory 70 m by 14 m is equipped with a total installed power load of 96 kW made up of the following: Line shaftings of two 18 kW slip-ring motors, individual machine drives for four 12 kW slip-ring motors, and six 2 kW squirrel cage motors.

Make a sketch plan approximately to scale showing the assumed positions of motors. Mark on the plan the electrical distribution you propose. Set out in schedule form the rated capacity and type of switchgear you would advise and size of conductors you would use (assume 0·85 p.f. and 85% efficiency for the motors). Ignore voltage drop. Nature of supply: 415-V 3-phase 50 Hz.

3. A large works contains a number of 415-V, 3-phase, 40 kW slip-ring induction motors. The supply is 6600 V 3-phase.

Describe with sketches or diagrams the lay-out of the necessary apparatus between the high voltage terminals and any one motor. [C]

4. The 3-phase 4-wire distribution system in a rural area consists of bare wires on overhead supports. Describe with sketches two different methods of giving a supply from the overhead line to a row of cottages, including the types of conductor used and the method of entry to the property. [C]

5. A 400-V, 3-phase, alternating-current electric supply is to be brought into the basement of a new block of flats. There are to be five floors in all and there will be three flats on each floor. A 60-A capacity, 240-V, single-phase supply is to be given to each flat and there are also a 4 kW lift motor and two 2·25 kW pump motors, all 3-phase machines, in the basement.

Describe the form of distribution you would install from the incoming supply to the main switches controlling supplies to the separate flats and for the motors, making sketches of any special ducts or openings that you would require provided in the building. [C]

6. The electrical installation in an office comprises 32, 100 W fixed lighting points, 8, 2 kW fixed electric heaters, and 16, 5 A socket outlets for office appliances.

The walls of the office are brickwork and the ceiling is of concrete supported on seven 0·3 m deep reinforced concrete beams, and there is a pillar at the centre of each beam. The installation is to be carried out in screwed conduit run on the surface and few from one main switch fixed at one end of the office. The supply is a 240-V, single-phase.

(a) Make a schedule of the wiring material and accessories that would be required, giving the size or capacity of all items including cable and conduit but omitting quantities.

(b) Draw a diagram to show the arrangement of each type of circuit to be run. [C]

7. A 2-wire distribution cable ABCDE, 575 m long and of uniform cross-sectional area, is supplied at both ends A and E at 400 V. Consumers' loads are connected as follows: 350 A at B, 175 m from A; 150 A at C, 425 m from A; 100 A at D, 100 m from E. The resistance of the cable, lead and return, is 0·1 per 1000 m. Determine:

(a) the direction and value of the current flowing in each section of the distributor;

(b) the point at which the load voltage is a minimum and its value at that point.

(c) the power loss in the cable. [C]

Chapter 8

WIRING TECHNIQUES

8.1. Fundamentals

It is often asked 'what is the best wiring system?'. There is no simple answer as the choice depends upon the technical and economic suitability for the conditions of service. The Code of Practice on Electrical Installations (CP 321 :1965) gives the following considerations as necessary for the design and planning of installations :

(1) The type of supply, and the earthing arrangements available.
(2) The probable maximum and minimum ambient air temperature in all parts of the installation.
(3) The possible presence of moisture or corrosive conditions on both.
(4) The possible presence of flammable or explosive dust, vapour or gas.
(5) The degree of mechanical protection necessary.
(6) The importance of continuity of service, including the possible need for standby supply for general or special purposes.
(7) The probable need for modification and re-wiring during the life of the building and the question whether it is necessary to avoid disturbance of surface finishes.
(8) The probability of future extension of the initial installation.
(9) The probable operating and maintenance costs, taking into account the electricity supply tariffs available.
(10) The relative cost of various alternative methods, e.g. of wiring, in relation to the estimated life of the installation.

Thus many factors have to be taken into account. Furthermore the type of wiring may be dependent upon as to whether the building is existing or under construction.

In a few years, installation work will have a history of 100 years and many methods for safeguarding against the effects of faulty conductors have come and gone. Passing reference is made to wood capping and casing—still used on the Continent with apparently few ill effects—where neatness required good carpentry ability. A survey of modern wiring systems is given below. Description is also given of newer and more specialised methods.

The general requirements for safety are contained in I.E.E. Regulations Ch. 13 Fundamental Requirements for Safety, being mostly extracts from the *Electricity Supply Regulations*, 1937 *and Electricity (Factories Act) Special Regulations*, 1908 *and* 1944. Since both are statutory, they form the compulsory section of the Wiring Regulations.

Cables are the life-blood of installations and recognised types are listed in the I.E.E. Tables 9D1–9N1. Any cable type sheathed with p.v.c., p.c.p. (polychloroprene), lead or having a h.o.f.r. (heat-resisting, oil-resisting and flame-retardant) sheath may incorporate a catenary wire where intended for suspension.

8.2. Conduit Work

In this country, conduit work is mainly carried out in heavy gauge (H.G.) screwed welded conduit to BS 4568, Class B; aluminium and copper conduits are also covered by this British Standard. The specification refers to brazed and close-joint light gauge (L.G.) conduit which is unthreaded and is often derisively termed 'tin-whistle'. In spite of this, the brazed type has been employed for the wiring of many thousands of flats, fulfilling the stringent requirements of the then L.C.C. (now the Greater London Council.) Where dry-conditions exist, close-joint conduit has been successfully used in small factories, stores and similar situations. Care is needed in bending to ensure that the tube-walls do not cave-in and such action may be avoided by the use of the bending machine. A band of approximately 15 mm at each conduit end must be properly filed and, after a push fit, tightly held by grip fittings to ensure continuity. All sharp edges will have to be reamered.

The screwed-conduit system is expensive and requires skill, experience and knowledge to be carried out properly. It may also entail heavy work, its popularity being almost entirely due to the strong mechanical, bonding and rewirable qualities.

To ensure a smooth internal bore, during manufacture any traces of irregularities in the welding are taken away by drawing the conduit through a mandrel. Removal of any rough internal edges by a purpose-made reamer (Fig. 8.1) rather than any old tool which comes to hand, is a basic requirement of good practice. Other essentials are:

1. All conduits must be completely fixed before wiring operations are commenced. This requirement does not apply to some flexible conduit systems as used for certain industrialised and prefabricated buildings (Reg. 521–11).

2. The maximum number of cables permitted for the various sizes of conduits was shown in I.E.E. Tables (14th Edition) B.5M and

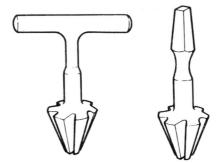

Fig. 8.1. Conduit reamers.

B.6M ; a relaxation was allowed for short straight runs, of which a switch drop would be an example, where a moderate increase may be made. It is not always realised that all cables must be *simultaneously* drawn in, otherwise, presumably, a larger size of conduit must be used. These tables require a great deal of attention especially when cables of varying cross-sectional areas have to be drawn into a single conduit. As an example, allowance was made for a maximum of 32 250/440 V p.v.c. cables in 32 mm conduit, but derating of the current-carrying capacity of such conductors would almost certainly be necessary. An indication of the method adopted for the I.E.E. 15th Edition is given in Example 8.1.

Cables must be drawn in straight without any twists and should not spiral off the coils. Wherever possible the cables should be run directly off cable drums (Fig. 8.2). They must be manually fed into the conduit and the action should be synchronised by the helper

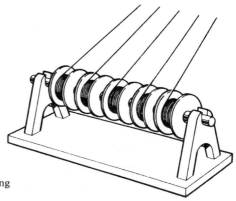

Fig. 8.2. Cable stand for feeding cables in conduit.

pulling at the other end; feeding is thus a two-handed operation. Not more than two 90° bends or their equivalent are permitted between boxes to allow for easy drawing-in, otherwise a reduction must be made in the number of cables installed. A little slack wire should always be left in draw-boxes; it facilitates re-wiring and relieves tension on the cables.

Example 8.1

In a conduit installation the length of run is 10 m. Assuming two right-angle bends, what is the conduit size to enclose four 2·5 mm² p.v.c. non-armoured cables?

From Table 12C, factor one 2·5 mm² cable = 30
So factor for four 2·5 mm² cables = 120
From Table 12D, suitable conduit size
with a factor of 141 is <u>20 mm</u>

COMMENT: These factors allow for conduit bends and presume easy cable pull into the conduit.

8.3. Continuity

Metal conduit practice must follow the basic principle that the system must be 'electrically and mechanically continuous'. The threads require to be thoroughly cleaned, as bad continuity in screwed joints arises from the presence of dirt, grease, rust or other poorly conducting surfaces between which contact must be made. Conduits should then be screwed in tightly without being over-turned. The type of connexion which operatives should aim at is shown in Fig. 8.3(a), it will be observed that the conduit is made to reach the shoulder inside the box-spout. There are no exposed threads, surplus threads being always a sign of bad workmanship.

Loose and floppy threads produce unnecessary electrical resistance. The looseness may also be caused by poor fixings resulting in weak mechanical supports. These points were brought out in a report on continuity.*

The report further states that loose joints may permit moisture or corrosive fumes to attack the contacting surfaces, and if subject to vibration may become even further loose and thus aggravate these

* The British Electrical and Allied Industries Research Association Technical Report F/T 157. 'Methods for Improving Electrical Continuity of Joints in Screwed Steel Conduit Systems' by E. E. Hutchings B.Sc. (Eng.), A.M.I.E.E.

Fig. 8.3(*a*). Tight connexions into boxes without any exposed threads.

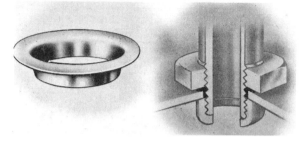

Fig. 8.3(*c*). Compression washers.

Fig. 8.3(*b*). Lead washers.

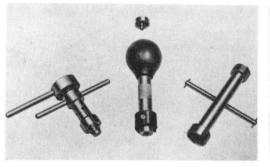

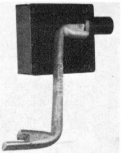

Fig. 8.3(*d*). Special tools for ensuring continuity.

effects. Stocks and dies should be held squarely and applied with an even pressure on both arms. Any swarf trapped in the teeth requires to be frequently cleared.

Trouble may arise where conduits have to fit into loop-in boxes, especially if there are three or more back-outlets, due to the inaccessibility of the joint and restriction of space for manipulation. The surface surrounding the holes may require scraping and the use of lead (Fig. 8.3(*b*)) or brass serrated earth-continuity washers. As an alternative, compression washers may be employed (Fig. 8.3(*c*)).

Special tools such as in Fig. 8.3(*d*) should form part of the electrician's kit for use on such occasions.

Running couplers ('runners') can also be a source of weakness, since once again they leave exposed threads; therefore in many cases the use of the *union* can make for improvement, incidentally saving the labour of cutting long threads. Figure 8.4 illustrates the two alternative methods.

To reduce the possibility of high resistance, the skilled craftsman will plan his runs so as to make for the minimum of couplers.

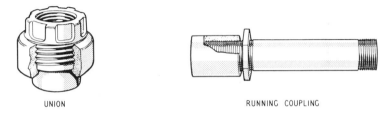

UNION   RUNNING COUPLING

Fig. 8.4. Conduit coupling methods.

8.4. Corrosion

Steel conduits under adverse conditions are particularly prone to corrosion hazard. High quality black enamel provides the first line of defence and therefore all imperfections which may be brought about by the actual operations, or from pipe-vice marks, should be made good by bituminous paint. Where there is dampness, galvanised conduit, which is a hot-dip process, gives protection against rust. The sherardised variety is an improvement in this respect, consisting of a zinc impregnation producing a dull-grey finish. Recently, conduit with a p.v.c. coating has become available.

Contact with the following class of materials can have a deleterious effect on conduits:

(1) Substances containing magnesium chloride which may be used in the construction of floors and dadoes.
(2) Plaster undercoats contaminated with corrosive salts.
(3) Lime cement and plaster (e.g. on unpainted walls).
(4) Oak and other acidic woods.
(5) Dissimilar metals liable to set up electrolytic action.

Condensation occurs where metal conduits are subject to changes in temperature. Warm air gives up its moisture as it strikes the cold walls of the conduit. Boxes filled with compound may be used to act as an effective barrier to the passage of moisture-laden atmos-

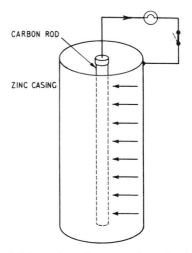

Fig. 8.5. Direction of current with a simple cell.

phere *through* the tubes. Small drainage holes in the boxes are sometimes advocated, as well as rubber gaskets under the box lids.

The Regulations stipulate that conduits should be self-ventilating, and drainage outlets must be provided at any points where condensed moisture might otherwise collect. One method of achieving this is by fitting an open spout at the lowest point of the conduit system. On building sites, it is advisable to swab out conduits before drawing in the cables.

CATHODIC PROTECTION. Metal sheaths of underground cables and metal pipes, when buried in certain soils, may corrode in a similar manner to the zinc casing of a primary cell (Fig. 8.5), where it will be seen that the electrode in which the current enters from the electrolyte does not corrode. *Cathodic protection* is one of the most important methods for preventing this type of metal sheath corrosion by causing a d.c. current to flow in the reverse direction (Fig. 8.6).

The anode consists of a block of magnesium or magnesium alloy which itself erodes away instead of the metal sheath, and the process takes a period of some 10/15 years. For this reason it is often referred to as a 'sacrificial anode'. The p.d. between the anode and the cathode being about 1 V.

Cathodic protection can be applied to steel, copper, lead, brass and aluminium in all soils.

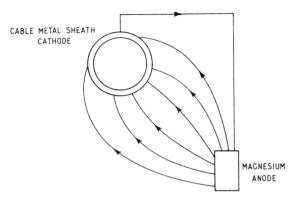

Fig. 8.6. Cathodic protection.

8.5. Boxes

The most commonly-used boxes are the malleable-iron types, misnamed B.E.S.A. boxes. The initials stood for British Engineering Standards Association, which for many years has been replaced by the British Standards Institution. Variations are many and a selection is indicated in Fig. 8.7(a).

These small boxes have a 60·3 mm diameter opening, and the 4M tapped holes in the fixing lugs are at 50·8 mm centres. Box depths are 25 mm and contrast with the large circular pattern of depth 30 mm. The larger types are designed with four tapped holes at fixing centres of 60 mm.

There is also a large range of rectangular cast-iron junction boxes with or without knockouts.

Modern concrete pre-cast block construction presents special problems to the installer. The loop-in angle box (Fig. 8.7(b)) avoids the necessity for a complete right angle and thus can be fitted where the floor depth is limited. Another manufacturer has produced a novel box as a further attempt to meet this problem (Fig. 8.7(c)), where it can be seen that no setting of the conduits whatsoever is necessary. Where cement mixtures are poured over the boxes, the joints must be almost watertight to prevent the ingress of liquid cement. On hardening, the cement can easily form an impenetrable barrier leading to costly replacement.

8.6. Further Conduit Precautions

(a) For vertical runs exceeding 5 m, clamping of the cables is necessary.

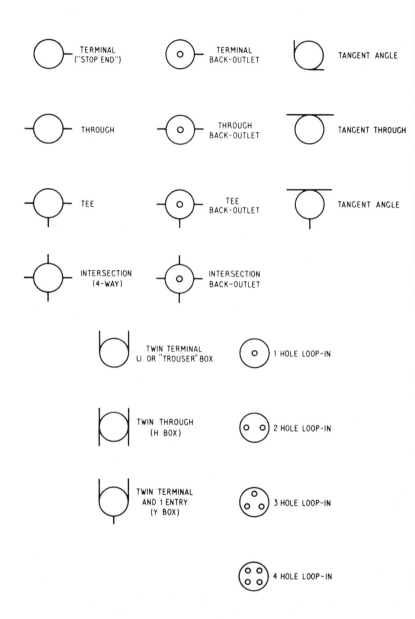

Fig. 8.7(*a*). Small circular boxes to BS 4568.

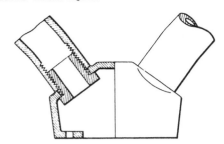

Fig. 8.7(b). Loop-in angle box.

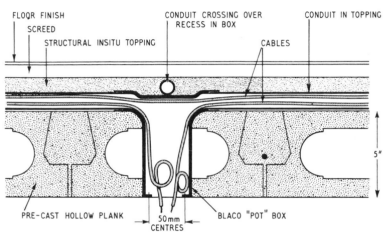

Fig. 8.7(c). Extension box for concrete "pot" floors.

(b) A single live conductor, fed by an a.c. supply, must not run by itself in conduit. Otherwise, due to the changing magnetic field cutting the conduit, eddy currents (Section 2.7) will be set up resulting in a heat rise. During a recent test, a current of 16 A was passed through an insulated conductor run in an isolated length of 20 mm conduit. An ammeter connected to each end of the conduit read as much as 2·5 A.

(c) With the need for *segregation*, low voltage and extra low voltage circuits, now termed Category 1 and Category 2 circuits respectively, must not be contained in the same conduit. This means, as an example, that normal bell wires, as fed from the household bell transformer, must not run together with cables supplied by the mains. There is an exception when the insulation of the extra-low-voltage cables complies with the requirements of the highest voltage present in Category 1. Fire-alarm circuits (Category 3) must in no circumstances be drawn into the same conduit with Category 1 cables.

(*d*) There is a great danger of floors being weakened by conduits being laid in floor joists. The depth of notches should not exceed that of the conduit diameter. Maximum strain occurs when the cuts are made across the centre of a room. Joists should therefore be slotted close to the bearings.

8.7. Plastic Conduits

In spite of the many advantages of heavy-gauge screwed conduit, as we have seen, the system has many drawbacks, notably tendency to corrosion and difficulties in obtaining and maintaining good continuity. Insulated conduit, also a draw-in system, suffers none of these disadvantages, yet it has to be admitted that widespread prejudice—probably based on earlier inferior types—has militated against its extended use and a wider knowledge of the undoubted possibilities. No such inhibition exists on the Continent; Holland uses nothing but plastic conduit and Italy, as an example, makes considerable use of it.

The conduits are highly corrosion-resistant to water, acids, alkalis and oxidising agents in the concentrations likely to occur in commercial and industrialised buildings. Mechanical strength is, of course, inferior to steel. Nevertheless a *high-impact* grade is available that is specially tough and durable, e.g. 20 mm conduit of this type will withstand an impact of 250 N falling on it from a height of 1·5 m.

An installation carried out with rigid p.v.c. conduits may appear to be remarkably like its metal counterpart. It would however be a mistake to slavishly follow the operations required for the traditional system. The basic possibilities and limitations must be taken into account. A British Standard for the materials is being prepared. At the moment the safe temperature range is from −15°C to 60°C although individual makers claim a higher maximum temperature.

Whilst threading of the rigid type may be carried out with new dies, or sometimes by the 'self-threaded' method of simply turning on the spout of a metal screwed box, the push fit is now generally favoured. This eliminates 'notching' which results in a drastic reduction of wall thickness. A solvent cement may be used at all joints and when the two parts are held together they become welded, being quite watertight in a few hours. A Bostic preparation can be used as an alternative and is preferable in most cases. Although providing a watertight joint, the preparation never sets hard but remains permanently tacky to allow for thermal expansion or future alterations. It should be borne in mind that the coefficient of thermal expansion of p.v.c. is roughly equivalent to an expansion of 6·5 mm in a 4 m length for a temperature rise of 45°C so that allowances may have to be made for this effect.

8.8. Plastic Conduit Installation Methods

A wide choice of insulated conduits are now on the market:
(1) Rigid super high-impact, heavy and light gauge.
(2) Flexible, heavy and light.
(3) Corrugated flexible.

The heaviest grade (high-impact type) may be able to withstand all site conditions even if the installation is left exposed to maltreatment by other services. This gauge is also suitable for the heavy type of industrial installation but requires mechanical protection if subject to heavy blows. The system is also suitable for chemical works, plating works and agricultural buildings. It may be used in tropical countries and where there is a consistently high temperature, within the limits already indicated, coupled with a humid atmosphere.

The high-impact light gauge is suitable for multi-storey flat installations, particularly if the tubing is fixed to the reinforcing mat and concrete poured within a few hours.

Obviously, the conduit does not form its own earth-continuity conductor and a green insulated wire of minimum size 2·5 mm² must be drawn in for this purpose.

BENDING. Setting of the rigid types of tube presents perhaps the biggest problem and no final solution has yet evolved. At the moment, several methods have been adopted.

Plumbers' bending springs. These coil springs are entered into the tube to prevent collapse at the bends. After setting, there is a tendency to revert to the original shape. It is therefore recommended to bend about 10° more than required, then bend back 5° for release before clearing the spring.

Bending is assisted by the slow warmth of a blow lamp, or preferably the yellow flame of a butane lamp. Various other methods may be used, such as a rag soaked in hot water, heated oil or sand bath.

Another method advocated is to warm over a spirit stove or lamp. After some 30 seconds the conduit will become quite flexible for forming into the desired shapes. If the bend tends to distort at the radius, stroke out gently but firmly with a rag or glove to perfect roundness. Hold in position until the conduit hardens in the ambient temperature.

For conduit sizes above 25 mm, solid rubber cord benders are adopted, since steel springs would leave corrugations in the plastic.

The polypropylene corrugated flexible type of conduit presents no bending problems and is of considerable interest. The writer has seen a complete site rapidly tubed by this method.

Due to the exceptionally smooth bore, cables are very easily drawn through most plastic conduits. In difficult cases, a smear of liquid paraffin should be applied to the cables.

WEIGHT AND HEAT. A non-metallic box of thermoplastic material must not be used for the suspension of totally-enclosed lighting fittings where the temperature of the box is likely to exceed 60°C. Further, the mass suspended from a non-metallic box is not to exceed 3 kg. Circular boxes can be supplied with brass inserts or special steel-insert clips to overcome this problem (Fig. 8.8).

Fig. 8.8. Steel clips are used to take the weight off plastic boxes.

8.9. Trunking

Normally composed of sheet steel, trunking is supplied in light or heavy gauge, of square or rectangular section, with detachable covers. Steel casing was first used in short lengths mainly at intake positions for linking switchgear to meters and to protect meter tail 'bites'. The position is now reversed in particular types of installation where the employment of conduits becomes subsidiary, more often being only required where two or more wires are needed to feed various points from the trunking.

Trunking has rapidly extended in order to become a system in its own right with modern improvements. The non-corrodible type, by zinc coating the mild sheet steel, would be employed in good class installations. There has grown up a multiplicity of fitments, tees, bends, elbows, etc. The latter are often avoided by a certain amount of ingenuity and the use of a portable welding set on site, but care must be taken that the cables do not suffer abrasion. The metal casings are relatively simple to install and have overwhelming advantages over the larger sizes of conduit in cost and the ability

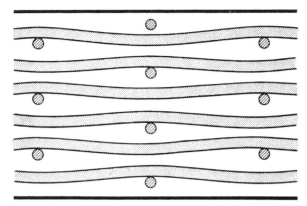

Fig. 8.9. Cables in trunking supported by a pin-type rack.

to tap off conduits at any point, by drilling holes or by means of a special punch.

Wiring is simplicity itself and is achieved by placing the cables in position. At 3 m intervals pin-type racks (Fig. 8.9) should be fitted in order to support the cables, the pins being covered with insulated sleeves. Spring-retaining cable clips may also be supplied for insertion as required.

For the illumination of large areas a common plan is to run parallel lines of trunking along the length of the building with perhaps fluorescent fittings suspended or fixed directly to the underside (Fig. 8.10). When batten-type fluorescent fittings are fixed 'head-to-toe' they form a continuous line of lighting.

With the phenomenal growth in the utilisation of electrical appliances in the last few years, installations are becoming more and more complex, bringing a corresponding increase in circuit wiring. Where a multiplicity of the cables are required to be housed, ducts and trunking appear to be the best solution we have to date.

A logical next step has been the extension into non-industrial situations. One example is the *skirting trunking* (Fig. 8.11), which is particularly suitable for commercial and, in some cases, domestic establishments. The partition as seen in the illustration is generally accepted as standard for this type of trunking and serves to segregate extra-low-voltage (here telephone) and low-voltage services.

Another common feature of this method is the provision of corner panels for easy access. Once the hollow skirting is installed in conjunction with the 13 A ring circuit it is a simple matter to add additional outlets. The existing cables can be used as draw-wires for any additional conductors.

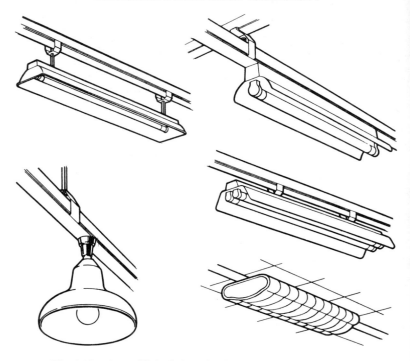

Fig. 8.10. Lines of light fittings fitted to the underside of trunking.

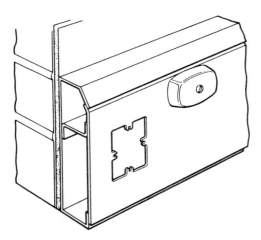

Fig. 8.11. Multicompartment skirting trunking.

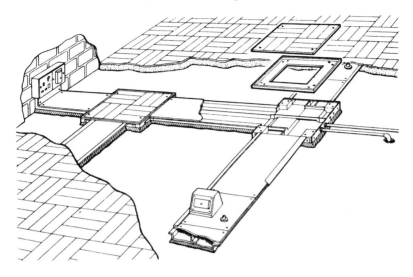

Fig. 8.12. Metal under-floor ducting system.

UNDER-FLOOR DUCTING. Neatness and ease of accessibility are now considered two major requirements of any newly planned wiring system. In older industrial installations, it is not unusual to come across an untidy mass of pipes covering the walls in all directions and often festooned between girders and ceiling trusses. More than likely, many of the conduit runs are no longer in use and completely out of service.

By contrast, the ducting system is not only invisible but permits almost immediate access and, because of this, circuit cables can be inspected for faults throughout their whole length or replaced in the shortest possible time.

Figure 8.12 illustrates part of an under-floor ducting system which normally extends into a simple grid (Fig. 8.13) arrangement so that tappings for accessories can be brought out for island positions. Ducting finds particular application in the larger offices with their continuously expanding use of electric business machines. The reader may see the arrangement in the laboratory of his technical college and the larger stores.

Particular features may be noted:

   (i) Partitioning for the segregation of the various electrical services.

  (ii) Conduit runs may feed the lighting of the floor below.

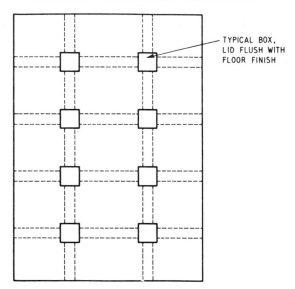

TYPICAL BOX,
LID FLUSH WITH
FLOOR FINISH

Fig. 8.13. Grid arrangement of under-floor ducting connected by intersection boxes.

(iii) The ducting may extend to the walls so that the wiring may continue along skirting trunking.
(iv) Metal ducting is electrically continuous so that bonding is automatic. The complete earthed enclosure acts as a barrier for the prevention of radio or television interference.

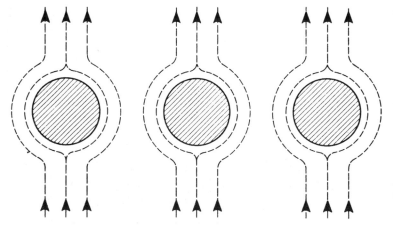

Fig. 8.14. Air currents circulating around cables.

Also available is an insulated type of ducting made of an impregnated fibrous substance.

Ducting demands a lower space factor than trunking or conduit. This reduced factor is due to the possible heat rise in the enclosed duct. The amount of electrical energy responsible for the heat is given by I^2Rt joules (I = current in amperes, R = resistance in ohms and t = time in seconds). However, the final cable temperature is also dependent upon the rate of dissipation of this heat. The generated heat would in the ideal case have been dissipated by self-induced air currents circulated in between and around the cable (Fig. 8.14) and along the inner surfaces of the ducting. The addition of extra cables produces a bulk which replaces an equal volume of air in addition to providing a further hindrance to the circulation of free air.

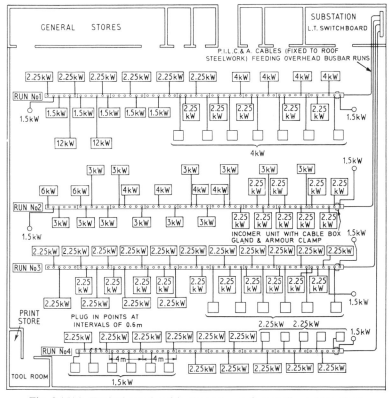

Fig. 8.15(a). Typical overhead busbar scheme for small machine shop.

8.10. Overhead-busbar Trunking

Overhead-busbar trunking forms yet another variation in the uses to which the basic rectangular sheet-steel arrangement can be applied. Since the solid copper or aluminium busbars can carry heavy loads, with tappings at frequent intervals, it may be seen as the power equivalent of under-floor ducting. Figure 8.15(*a*) shows a typical arrangement for a machine shop, while Fig. 8.15(*b*) gives alternative types of trunking feed boxes.

Students of the history of wiring will have noted that in the early period fuse protective devices were placed near to the consuming apparatus—consisting mainly of lamps. It was common to find fuses scattered all over the premises, in switches, ceiling roses, under floor boards and in almost any inaccessible position, bringing consequent difficulty of location when a fuse blew.

The same principle has been adopted for factory busbar distribution, where the fuses are once again located near the consumer devices

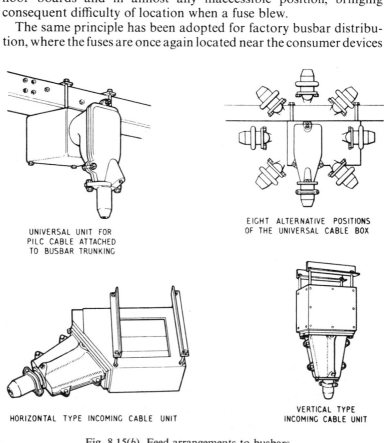

UNIVERSAL UNIT FOR
PILC CABLE ATTACHED
TO BUSBAR TRUNKING

EIGHT ALTERNATIVE POSITIONS
OF THE UNIVERSAL CABLE BOX

HORIZONTAL TYPE INCOMING CABLE UNIT

VERTICAL TYPE
INCOMING CABLE UNIT

Fig. 8.15(*b*). Feed arrangements to busbars.

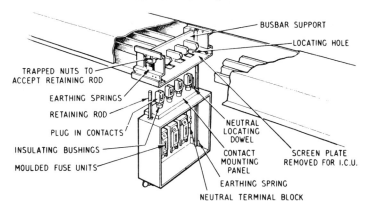

Fig. 8.16. Plug-in type fusebox.

or apparatus, but with accessibility very much in mind. What has brought about this somersault? Clearly the very multiplicity of electrical appliances has made the need for decentralisation.

Tapping points are situated along the overhead busbar trunking at 0·61 m intervals where the busbars can be connected to a plug-in type box (Fig. 8.16) which contains the necessary fuses for sub-fusing protection of the motors. This scheme allows for the maximum flexibility where rearrangement of the shop floor is required. The fused plug-in boxes can be plugged in or withdrawn with perfect safety while the busbars are alive. The overhead busbar trunking is usually fitted at roof-truss height. All floor machines are fed from the boxes of heavy-gauge screwed conduit.

The bars may be obtained in a variety of current ratings from 60 A to 1500 A with fuses ranging from 15 to 500 A. Sometimes when the trunking is used as a ring-main, it is fed by separate supplies at opposite points; in the event of one supply breakdown, the ring can still be fed from the other supply.

8.11. Mineral-insulated Cables

The mineral-insulated cable is probably the simplest of all types, being made up from two materials only—copper for the sheath and for the unstranded conductor(s) and tightly-packed magnesium oxide (MgO) powder for the insulation. An alternative cable with aluminium sheath and conductors is available but, as with cables of every other type, aluminium conductors of cross-sectional area 16 mm² or less are not yet recognised by the I.E.E. Regulations.

The same tools must not be used for aluminium and copper cables as minute traces of copper can promote corrosion due to electrolytic action.

The major advantage of m.i. cables is considered to be their ability to withstand high temperatures without deterioration. They can operate quite normally up to 250°C although with a p.v.c. oversheath the limiting temperature is 80°C. Standard terminating seals must not be allowed to reach more than 150°C. For continuous operating temperatures of up to 150°C, medium temperature seals have been introduced, consisting of fibre-glass discs and chemically-inert fluorinated propylene (f.e.p.) sleeves. Special high temperature compound must be used.

I.E.E. Tables 9J1–9J6 give the current ratings for the 660-, 440- and 250-volt grades. It should be noted that while the current rating is higher than, for example, equivalent p.v.c.-insulated cables, the permitted volt drop is reduced and this factor must be taken into account when making calculations.

Because the mineral insulation is hygroscopic (i.e. it absorbs moisture) sealing of all cable ends is required as shown in Fig. 8·17. The pot is self-threading and when carrying out this operation the use of the pot wrench is of advantage as it ensures the pot being squarely threaded on to the cable. The sheath must not protrude into the pot but should be made approximately level with the shoulder. Sealing by the wedge type of pot simplifies the procedure.

The purpose of the gland is to allow for threaded connexions to metal boxes. Bonding is effected by the compression ring, or 'olive', which shrinks on to the cable sheath as the gland nut is tightened.

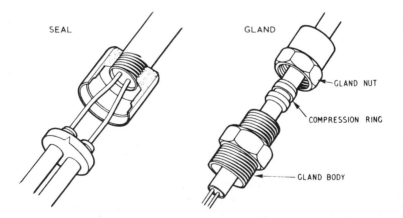

SEAL   GLAND

GLAND NUT

COMPRESSION RING

GLAND BODY

Fig. 8.17. Mineral-insulated cable termination.

For competitive work a simpler type gland with a grub screw for bonding is often employed. Plaster-depth boxes may be obtained with special clamps that obviate entirely the need for glands. Stripping of the sheath is facilitated by the rotary stripper and for long lengths there is a motor-operated type, but care is still required for this action. Operators responsible for installing mineral-insulated cables should make themselves familiar with manufacturers' instructional booklets and themselves analyse their own actions and seek improved methods. Some of the requirements for assuring good practice are given as follows:

1. Operator's hands should be washed before making the seal, since absolute cleanliness is essential. Metal particles or other foreign bodies will, if present in the seal, lead to a breakdown. After screwing on the pot and before filling with compound, the careful worker will quickly examine the pot interior. The use of insufficient compound is a false economy, filling should take place from one side to avoid the possibilities of air pockets. After the seal has been made all surplus compound should be wiped away.

2. On no account must there be any delay in completing the seal once the cable has been stripped, otherwise the insulation value will rapidly fall with the absorption of moisture and the dampness may be retained permanently in the cable. Puncturing of the sheath, even to the extent of a pin-hole, also results in the absorption of moisture from even slightly humid atmospheric conditions. During erection, care will obviously be taken to avoid piercing by sharp instruments, but in addition when large coils are delivered to the job the hessian or other protective wrapping should be retained on the cable until it is actually fitted. Similarly, it is advisable to retain pre-assembled cable units in their boxes right up to the last possible moment. On no account should the cables be laid on sharp stones. For concealed work cables are normally run in concrete floors. The keen electrician will zealously guard his cable until pouring takes place and the cement is set. Exposed m.i. cables rising from the ground must be protected to a height of at least one metre with metal channelling or bushed conduit.

3. The present rule for bending is: *the minimum internal radius bend must not be less than six times the overall diameter of the cable sheath* (*Fig.* 8.18). A point to notice is that cable *work hardens* on bending. For subsequent straightening and re-bending, it may be advisable first to re-soften the copper sheath by bringing it to red heat with a blow lamp or blow torch.

4. To prevent the spreading of fire or smoke by the cable passing through floors, walls, partitions or ceilings, the holes must be made

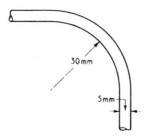

Fig. 8.18. An example of the minimum radius required for mineral insulated cable.

good with cement or similar fire-resisting material to the full thickness of the floor, wall, etc. Where threaded through holes in structural ironwork, all such holes must be bushed to prevent abrasion of the sheathing.

5. Copper can suffer attack through chemical or electrolytic action and wherever such dangers exist a p.v.c. oversheath is required. Where the p.v.c.-covered cable is terminated, the oversheath is stripped back and inevitably a short length of the copper sheath is left exposed. To minimise the possibility of corrosion the use of a corrosion-resistant hood or p.v.c. shroud should be fitted to overcome this problem.

Localised protection against corrosive effects may be given to short lengths of cable by lapping with adhesive p.v.c. tape. It is particularly important to do this where the cable emerges from a concrete floor, since corrosive elements are likely to be washed down and accumulate around the cable. To prevent any trouble, apply the p.v.c. tape (before the floor has been laid) extending to about 150 mm above and below floor level.

Bare cables should never be buried under a terrazzo finish as these floors may be polished with hydrochloric acid, followed by ammonia. The acids are absorbed by the concrete and the cable will become corroded.

Certain plasters and cement with an acid content also corrode copper. Mildly alkaline plasters are not corrosive provided they are free from ammonia; lime plasters and portland cement are quite safe to use. For further advice on precautions against corrosion the reader is directed to Appendix 10 of the I.E.E. Wiring Regulations.

6. Magnesium-oxide insulated cable sizes from 25 mm² upwards are only obtained in the form of single core. Where several of these larger sizes make up a run in parallel, care must be taken to neutralise the inductive effects which may be manifested as *eddy currents* or sheath *circulating currents*. The cables should not be spaced apart but should run in close proximity. Where runs comprise three cables,

the trefoil (triangular) formation is advisable in order to reduce the heat rise and losses.

For overhead lines it may be necessary to offer protection against the effects of voltage surges induced by lightning. *Surge diverters* are connected between the conductors and earth at the incoming supply terminals. A surge diverter is a form of *non-linear resistor*, which is practically an insulator at mains voltage and becomes increasingly conductive as the voltage rises. The diverter thus provides an alternative parallel path for any surge voltage.

8.12. Earth Concentric Wiring

Since the sheath can be used as an external return conductor, mineral-insulated cables admirably lend themselves to this system. There is an obvious gain in replacing two cores by a single core for normal wiring purposes, whilst three phase and neutral may be supplied by 3-core cables. In addition, since the copper sheath has a much lower resistance than the internal conductor for all sizes up to 10 mm², there is a reduction in voltage drop for any given length of run. To cut costs even further suitable *backless* boxes have been produced.

Earthed concentric wiring, however, is only permitted where the Electricity Board has been specially authorised to allow additional connections to the neutral conductor to earth (normally by protective multiple earthing). Otherwise the supply must be obtained via a transformer or converter, so avoiding metallic connexion with the public supply. Alternatively, a private generating plant may be employed.

For earth-concentric wiring purposes, the sealing pots are provided with earthed tails, crimped to the inside of the pot, so that the cable sheath now becomes the combined earth-neutral conductor. At switch and outlet terminations, bonding may require special attention. As the sheath carries the full current, reliance cannot be placed simply on clamping and, in this respect, copper-alloy boxes have an advantage over cast-iron or sheet-steel types.

8.13. Plastic Cable Wiring Methods

Practically all private houses are now wired with p.v.c. *insulated and sheathed cables* conforming to BS 2004 : 1961 and subsequent amendments. The p.v.c. compounds used are non-hygroscopic, tough, durable, resistant to corrosion and chemically inert.

In new work, the 3-plate method for the connecting lighting points offers advantages since all connexions can be made at the second fixing stage. However, the method is rather wasteful of cable so that in existing buildings the use of plastic joint-boxes is preferable as it also avoids the grouping of cables at outlet positions.

For two-way light control, a twin-and-earth cable may be brought to one switch position from which a 3-core and earth cable is linked with the other switch. This method is also wasteful and may entail cutting away of brickwork. Again, the use of a joint-box, as shown in Fig. 8.19, can effect savings in this respect. The illustration also shows the requirement of an *earth terminal at each switch and outlet position*. It is important to note that identification of cable cores requires a red sleeve for the black wire of twin-and-earth cables at switch positions.

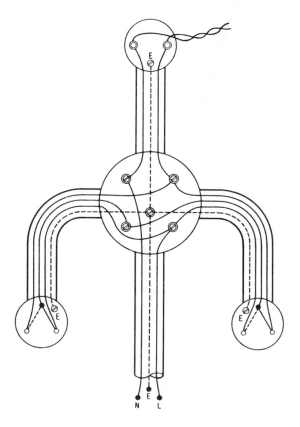

Fig. 8.19. Two-way light control with p.v.c. sheathed casing.
The joint-box avoids doubling back at one switch position.

In view of the growing number of electrical appliances being made available, a liberal supply of socket-outlets is recommended. In this way the amateur wiring enthusiast may be less tempted to carry out unauthorised extensions and the use of multiplug-adaptors or long trailing flexes would be minimised. There is little doubt that a fundamentally-safe installation can be rendered dangerous if extended by these means. The actual position of outlets is also important, so that the outlets are not concealed behind heavy furniture or allow flexible cords to cross doorways. The bottom of the socket-outlet plate should not be less than 150 mm from the floor level. This minimises the possibility of the outlet plate being kicked and also avoids undue bending of the flexible.

PROTECTION. There must be adequate protection against mechanical damage. Installing the cables under timber floors requires that the joists should be drilled. The tops of the holes should be at least 50 mm from the underside of the floor boards. Melting of the sheaths by frictional heat can occur if the cables are drawn too sharply through the holes or across each other. An application of light grease or Vaseline will ease this drawing-in process.

The minimum radius of bending for cables is given in I.E.E. Table 52C.

FIXING. For sheathed cables run under flooring where they are not likely to be disturbed and, provided that the surface is dry and reasonably smooth, additional supports are not required. Where the cables are exposed to view, the spacing distances for fixing, as given in I.E.E. Table 11A, should be maintained.

As a guide, the maximum overall diameter of certain p.v.c. twin-and-earth sheathed cables may be taken as 8·3 mm for 1·5 mm². 9·7 mm for 2·5 mm² and 12 mm for 4 mm².

8.14. Polychloroprene-sheathed Cable

Electrical installations in farm yards are subjected to severely onerous conditions. Danger to cables arises from the mineral acids and alkalis present; also steam, ammonia, sulphur fumes and lactic acid all play their part in attacking normal wiring. A special farm-yard cable has been developed of tough-rubber-sheathed (t.r.s.), taped, braided and compounded construction. Alternatively, the wiring can be carried out with polychloroprene (p.c.p.) sheathed cables.

P.c.p. is a rubber-like material and is also known as *neoprene*. The insulation is proof against the highly-corrosive farm products and also against weathering. Ozone, which is formed by the action of the sunlight's ultra-violet rays, attacks exposed natural rubber, whilst p.c.p. cables are unaffected by this action. The sheathing may with advantage be coloured black for outdoor use. All-insulated moisture-resistant accessories should be fitted.

8.15. Heat-resisting Cables and Flexibles

One of the newer problems, which the present generation of installation workers have to face, is the effect on plastic cable insulation of subjecting it to increasing temperature rises. For this reason, a number of special synthetic insulating materials have been developed for cables and more can be expected. Whilst ordinary p.v.c. has been found to be satisfactory for general wiring, cable research has brought out materials with better heat-ageing qualities, such as heat-resisting p.v.c. and chlorosulphonated polythene (c.s.p.) for sheathing purposes Nitrile butadiene rubber/p.v.c. (n.b.r./p.v.c.) possesses oil resistance superior to that of older types of cable. Such specialised cables now come under the heading of h.o.f.r. (heat-resisting, oil-resisting and flame-retardant) sheath complying with Part 4 of BS 2899.

The term *elastomer* is often used in connexion with this class of materials since the chemical compounds require to be vulcanised or cured so as to convert them to a suitable tough and elastic condition. This process has been applied to natural rubber, p.c.p., butyl rubber, silicone rubber and to chlorosulphonated polyethylene. A very recent addition is ethylene propylene rubber (e.p.r.) which is stated to possess electrical properties superior to butyl rubber.

Chlorosulphonated polythene (c.s.p.) as a raw material is similar to polythene with chlorine and sulphur incorporated in the molecular stage. As a cable it may go under the trade name of 'Hypalon' and has a lower water absorption than polychloroprene. It does not discolour as easily as p.c.p. when exposed to sunlight. The operating temperatures, as given by the manufacturers, are: $-30°C$ to $85°C$ for flexibles and $-40°C$ to $85°C$ for cables.

In order to increase efficiency, filament lamps are being run at continuously higher temperatures. Some of the temperatures reached are shown in Fig. 8.20 for a 100 W coiled-coil lamp, and require a heat-resisting type lampholder where it is likely to attain a temperature in excess of 135°C. The use of heat-resisting lampholders is likely to be necessary where the lamp will be installed in the 'cap-up' position or where the lampholder is contained in an enclosed lighting fitting or an inadequately ventilated shade.

Referring to Fig. 8.20, at these temperatures p.v.c. is completely

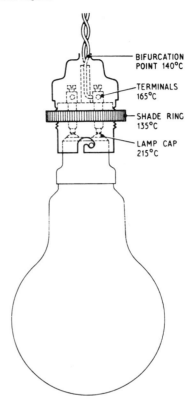

BIFURCATION
POINT 140°C

TERMINALS
165°C

SHADE RING
135°C

LAMP CAP
215°C

Fig. 8.20. Temperatures reached at a
100 W coiled-coil lamp.

unsuitable and rubber insulation becomes brittle; the slightest disturbance may crack the insulation, leaving exposed bare conductors. For flexible cords, the Electric Cable Manufacturers' Confederation (formerly, the Cable Makers Association or C.M.A.) recommends the use of silicone-rubber or varnished glass-fibre insulation as the most suitable types for all-round use. Each conductor of the latter is wrapped with two layers of heat-resisting glass fibre which are impregnated with a suitable varnish. A braid of similar fibre is applied over the primary insulation and impregnated with varnish.

Where connexion is to be made to a batten-holder, the fixed wiring should terminate at an adjacent joint box from which a heat-resisting flexible is then taken to the batten-holder. Another method is to slip heat-resisting sleeving over the individual cores of the fixed wiring.

Butyl rubber, made up circular with cotton filling, cotton-braided and white flame-retardant cellulose-lacquered, is often used for the connexion to apparatus, such as immersion heaters, where the terminals are at an elevated temperature.

To comply with the Regulations, lampholders must be marked 'BS 52A'. Where a holder is likely to attain a temperature in excess of 135°C it must be of the heat-resisting type and marked 'BS 52H'.

8.16. Paper-insulated and Other Power Cables

Paper-insulated lead-sheathed cables are mostly used for underground feeders and distributors. Due to the high current and voltage rating, the industrial electrician may have to be familiar with their use.

A general view of a 1100-V 4-core p.l.s.w.s. (paper-insulated lead-sheathed served, single-wire armouring with overall serving) may be seen in Fig. 8.21.

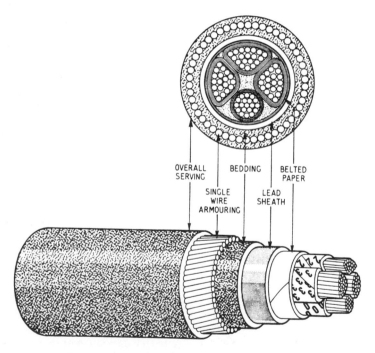

OVERALL SERVING   BEDDING   BELTED PAPER

SINGLE WIRE ARMOURING   LEAD SHEATH

Fig. 8.21. 1100 V 4-core paper-insulated lead-sheathed single wire armoured cable.

The conductors may be of copper or aluminium (stranded or solid). In the illustration they are of high-conductivity plain copper wires insulated with many layers of long-fibred oil-impregnated papers specially selected for strength and chemical purity. The identification of the cores is in accordance with the cable standard, BS 480: 1954, in which the numbers 1, 2 and 3 indicate live conductors, and 0 indicates the neutral. In order to distinguish between 1100-V cables and higher voltage cables, all 1100-V cable includes a yellow varnished paper tape applied longitudinally under the bedding.

The cores are laid up together and wormed circular. The worming, which fills up the space between the core and belt papers, consists of jute stringing or paper fillers. The belting tapes are also impregnated with chemically-pure hydrocarbon oil and refined rosin.

The continuously-extruded lead sheath is designed to keep out moisture, which would have a damaging effect on the insulation. Alloying increases the fatigue resistance of pure lead; therefore the addition of suitable metals such as tin and antimony makes the sheathing more suitable to situations where vibration is likely to be encountered. The lead sheath is, in addition, used for earthing purposes.

Protection may be afforded by armouring, consisting of two coats of steel tape applied helically in the same direction. This is particularly suitable for cables laid directly in the ground. When pulled through ducts or subject to bending, galvanised steel wire armouring is preferred. The cable may be made of two or more layers of wire armouring according to the degree of protection required.

Serving or bedding lies between the lead sheath and armouring and usually consists of compounded paper together with layers of yarn, jute, bitumen, cotton or hessian tape. Serving also prevents the armouring from cutting into the lead sheath.

The wire armouring may complete the cable in which case it is said to be 'left bright'. A serving finish acts as a protection against oxidation and corrosion and again may consist of paper, cotton and hessian tapes treated with a bituminous compound.

For vertical runs the fluid oil tends to run. The draining may leak into boxes and switchgear. To overcome this disadvantage mass-impregnated non-draining (m.i.n.d.) cables have been developed. The non-draining properties are produced by impregnating the oil with special non-draining high-melting-point waxes.

8.17. Laying P.I.L.C. Cables

Paper-insulated cables may be buried direct in the ground at a minimum depth of 460 mm and protected by earthenware tiles.

They may also be drawn through ducts, but here the current rating is reduced.

Bending sweeps should be as easy as possible. The minimum internal bend radius should be $12 \times D$ where D is the overall diameter of the cable.

For smaller sizes, plain galvanised saddles may be used for fixing at the maximum distances shown in Table 11A of the I.E.E. Regulations. Open- or claw-type heavy cleats of cable hangers are employed for bigger cables. Cable hangers have the advantage of ease of installation and allowance for additional cabling. Recently, cable trays designed for the fixing of these cables have been made available.

When off-loading, the drum may be easily damaged and must not be dropped, otherwise flattening of the cables will result, particularly if unarmoured. If there is no vehicle crane, the drum must be run off the lorry down stout planks, the drum itself being held under control by means of ropes attached to the spindle.

At the correct position, the drum should be set up on jacks and cable taken from the top, the cable being braked to avoid over-running; the cable end is pulled in the direction of the arrow as indicated on the drum.

Ample plant and labour are required to avoid cable strain. When unrolling, the natural sweep of the bend should be maintained and plenty of light grease or petroleum jelly applied when pulling the cable through ducts. At all costs, twisting must be avoided and rollers must be used to prevent abrasion.

For pulling the cable there are two kinds of grip, namely the stocking pulling grip and a pulling eye attached to the conductors.

8.18. Jointing Paper-insulated Cables

Jointing of paper-insulated cables requires special skill, practice and experience. Consideration of the cable make-up, as already explained, is of considerable help. Basically, the process is to safely expose sufficient length of the conductor for jointing. After the technique of joining has been carried out, the insulation is to be remade equal to the original and then covered for full protection.

The actual jointing process for a through-joint may be summarised as follows:

1. The cables are cut to give about 75 mm overlap and a lead sleeve is run over one cable. A wire binder or 'snugger wire' is then wound on the cable where the wire is to be stripped according to the box dimensions.

2. If the cable is served it should be stripped back to the snugger

wire by means of a blow-lamp to assist removal. The armour wires are partly cut through and snapped off by bending backwards and fowards. A guarded blade is preferable so as to prevent any possibility of cutting into the lead sheath.

3. The sheath must be cut back most carefully by means of a glazier's hack-knife, or similar tool, allowing 25 mm each side of the sleeve at both ends. A light cut is made so as not to pierce the lead, completely round the cable. From this circle a cut is made to the end, allowing the lead sheath to be peeled off. The cut end of the sheath is then bell-mouthed in order to prevent the possibility of it cutting into the paper insulators.

4. A temporary linen tape is tied round the belt paper close to the cut sheath. Surplus belting and filler can now be removed. Another whipping of tape is placed over each core to allow the safe removal of the core papers by means of a sharp knife. These core papers are stripped back for a distance equal to just over half the length of the weak-back jointing ferrule.

5. The cores, which are now cut to approximately 6 mm overlap, are set in position and held in place by means of wooden wedge-shaped spreaders. Tinning the conductors is carried out by the pot and ladle method using resin acid-free flux. The joint is now made by sweating the cable ends on to the ferrule (Fig. 8.22). Whilst the solder is molten the ferrule is closed with gas pliers. All surplus solder should now be wiped off; using a small mirror for viewing inaccessible parts. A point to note is that the solder is at the right heat when white paper dipped into it turns brown.

6. The temporary tapes are removed and moisture-free linen tape is used to insulate the joint, half-wound lappings being made to

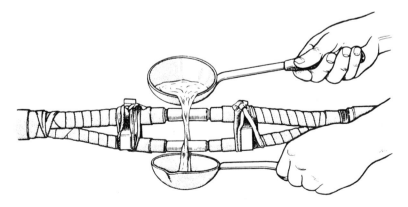

Fig. 8.22. A through-type joint on paper-insulated cables using the pot and ladle.

about $1\frac{1}{2}$ times the original insulation thickness. The spreaders are removed and the sleeve is plumbed on to the sheathing by means of wiped joints.

7. The sleeve is filled with bitumen compound heated to manufacturer's instructions. Allowance should be made for cooling before topping up.

8. The outer cast-iron box is then bolted in position and a further filling with compound completes the joint.

A tee-joint follows a somewhat similar procedure except that claw-connectors may be used for the actual joint. A split lead cover replaces the lead sleeve. Aluminium conductors require a special flux for jointing; compression tools are also used. The removal of an aluminium sheath is more difficult than a comparable lead sheath.

8.19. P.V.C. Armoured Cable

Particularly for indoor applications, p.v.c.-armoured cable has proved a serious competitor to the paper-insulated counterpart. This cable is tough, tighter and more flexible. In fact, paper cables only score through having a somewhat wider temperature range and higher current rating. The p.v.c. temperature range is 0°–70°C; below 0°C cracking of the insulation can occur. The respective maximum current-carrying capacities are: p.v.c.—480 A, paper insulated—570 A.

P.v.c. cables to BS 346: 1961 are simple in construction. There is a p.v.c. bedding for the wire armour over the p.v.c.-core insulation, and the cable is served with a p.v.c. overall sheath. For single-core cables the armouring would be non-magnetic.

Aluminium conductors are used extensively and may be of stranded or solid type. Above 25 mm aluminium becomes a more económic proposition than copper.

Probably the major advantage lies in the simplicity in jointing. Sweating is no longer required; lugs can be fitted by compression tools, which are also being used for through and tee joints at joint boxes. In this respect it might be mentioned that bitumen compound is replaced by epoxy resin.

Terminations require simple mechanical cable glands (Fig. 8.23) which bond and anchor the armour wires besides providing a damp-proof and dustproof assembly. The addition of a sealing ring renders the device suitable for external work. Care must be taken in selecting the appropriate size of gland and, due to cable tolerances, it might be advisable to determine the actual diameter of the cable with which the glands are to be used.

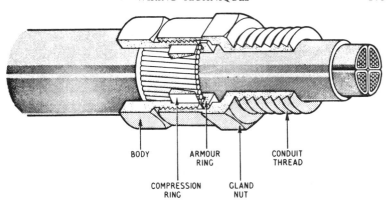

BODY ARMOUR CONDUIT
RING THREAD

COMPRESSION GLAND
RING NUT

Fig. 8.23. P.V.C. armoured cable gland.

8.20. New Wiring Systems

There has been a large number of developments mainly with regard to the need for wiring techniques that can be adapted for use in industrialised or prefabricated buildings. Here, the traditional piecemeal methods of installing wiring are no longer suitable. Building units consisting of complete walls and ceilings containing pre-cut and pre-assembled wiring are fitted together in accordance with tight site schedules.

In industrialised buildings, electrical installation materials become more amenable to precise calculations and they have even been brought out in a packaged form as part of an engineering service. It is instructive to quote the relaxation permitted in I.E.E. Regulation 521–10 'The conduits for each circuit shall be completely erected before any cable is drawn in. This requirement does not apply to prefabricated flexible conduit systems which are not wired *in situ*'. However, adequate allowance must be made for variations in building dimensions so as not to put any strain on the wiring. Also proper precautions must be taken against damage to any conduits or cables.

The flexibility of p.v.c. cables is used to the full and cable connexions between wall, floor and ceiling units may be made by suitably-insulated plug-in spade terminals.

The ideas produced by the need for mass-produced installations are bound to demand new approaches to standard wiring. Already, there is a simplified conduit system which saves the burden of pipe-threading by using threaded inserts which are knocked on to light gauge aluminium conduit. Rising mains can be made much more compact by means of the new co-axial riser. An interesting innovation from America is the *stretched cable*, which permits repeated cable extensions to three times its original length, and finds its application for light-current instruments and appliances.

8.21. Flameproof Equipment

The flameproofing of equipment is necessary to guard against special industrial risks. Regulation 27 of the ELECTRICITY (FACTORIES) SPECIAL REGULATIONS, 1908 and 1944, states:

'All conductors and apparatus exposed to the weather, wet corrosion, *inflammable surroundings or explosive atmosphere*, or used in any process or for any purpose other than for lighting and power, shall be so constructed, or protected, and such special precautions shall be taken as may be necessary adequately to prevent danger in view of such exposure or use.'

Flameproof gear and wiring are essential if there is extra risk of fire and explosion. Such hazards may arise in certain quarries, in coalmining, and in the chemical, paint and petroleum industries.

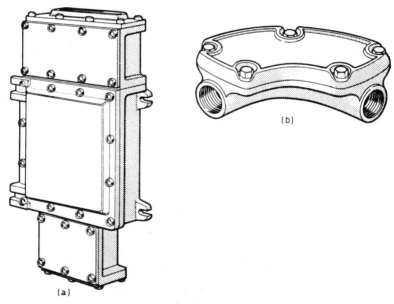

Fig. 8.24. (*a*) Flame-proof distribution fuseboard. (*b*) Flame-proof bend.

Thus, flameproof equipment must be installed for the wiring of petrol pumps and cellulose car spraying. Certified flameproof electrical apparatus and fittings (Fig. 8.24) will prevent the transmission of flame such as will ignite any flammable gas. It bears the mark registered by the Department of Energy together with the serial number of the Certificate relating to its type. The emblem takes the form of the outline of a crown containing the letters FLP. As may be observed, flameproof fitments are of massive cast-iron construction with flanged-type covers to withstand possible explosion. The flame of such an explosion is cooled by its passage through minute flange gaps.

8.22. Flameproof Wiring

Approved systems are *solid-drawn* screwed conduit, mineral-insulated cables and armoured cables. For the screwed-conduit method, 25-mm long minimum threads are required, as are stopper boxes where the conduit run enters non-flammable areas. These boxes contain a barrier to prevent the transmission of explosive gases. Mineral-insulated cable glands must incorporate a longer threaded body. Armoured p.v.c. cable with a p.v.c. sheath is considered as flameproof except when it may be contaminated by flammable volatile liquids if buried underground.

Intrinsically safe. This is a term that is applied to apparatus by which electric sparking in gases is incapable of causing an explosion.

Pressurisation. Sometimes security against the entry of explosive gases in a hazardous area may be achieved by passing slight air, CO_2 or inert gas pressure to the interior of fittings.

TABLE 9.1. CLASSIFICATION OF EXPLOSIVE GASES

Group no.	
I	Gas encountered in coal mining—methane (firedamp)
II	Various gases met in industry, such as cellulose vapour, petrol, benzine, amyl acetate, etc.
III	Coal and coke gas and ethylene oxide.
IV	Excluded gases, i.e. where there is no flameproof general approval, such as for acetylene, carbon disulphide and hydrogen, although approval may be given in individual cases.
	The full list is given in BS 229.

EXERCISES

1. Part 1 of I.E.E. Regulations states: "Good workmanship is essential for compliance with these regulations."

Give your definition of compliance in the following cases:

(a) treatment of cables and wire at points of connexion to control gear, distribution boards and bus-bars;

(b) treatment of heavy gauge conduits before and after erection;

(c) means of protection against spread of fire where conductors pass through walls and floors and where they are installed in vertical cable ducts;

(d) condensation in conduits;

(e) installation of cables feeding buildings at a distance greater than 3 m apart. [C] as amended

2. Describe briefly the systems of wiring you would adopt for the following installations and conditions, giving your reasons for selection: (a) boiler-house with high ambient temperature, (b) a petrol store, (c) a farm, (d) a factory with a large number of portable electric tools. [C]

3. (a) Describe and illustrate with a sketch a typical terminal connexion for a 2-core mineral-insulated cable.

(b) Explain why the current rating for this type of cable is higher than that of p.v.c. cables.

(c) Give examples of conditions where in your opinion the use of this type of cable is preferable to p.v.c. cable in screwed conduit, and give your reasons. [C] as modified

4. You are given the task of off-loading from a truck an electric motor having a mass of 1300 kg, transporting it to a working position and lifting on to a 2 m high platform with no crane available. A nearby builder's yard has available planks, poles, cords, block and fall, rollers.

Describe the procedure you would adopt in tackling this operation. [C]

5. Describe for the following installations the precautions you would consider necessary to prevent the risk of spread of fire.

(a) Vertical bus-bars rising through a multi-storey building in specially-formed duct.

(b) Oil-cooled transformers.

(c) Electrical equipment in a paint-spraying shop. [C]

6. In supervising the installation of a system using p.v.c. cables in steel trunking what would you consider to be the essentials to ensure sound construction, long life, and compliance with the I.E.E. Regulations?

[C] as modified

7. Describe the type of equipment and the methods of wiring which should be used to comply with the I.E.E. Regulations for the installation of a supply for each of the following on a building site:

(i) a stationary continental tower crane of total load 30 kW, 415-V, 3-phase,

(ii) four 240-V, 1500-W tungsten-halogen floodlights mounted on the crane,

(iii) four points in the building, each to consist of three socket-outlets for portable tools,

(iv) temporary lighting in a number of small basement rooms,

(v) a concrete batching plant of load 60 kVA at 415 V, 3-phase.

The available 3-phase and neutral supply is 415/240 V.

Summarise the complete installation by means of a single line schematic diagram giving the types of cable and protective device. Show the voltages at the above load points. [C]

8. (a) Describe the method of making-off the end of a 2-core mineral-insulated metal-sheathed cable, freely illustrating your answer with sketches of each stage of the task.

(b) Comment briefly on the following considerations in the use of mineral-insulated metal-sheathed cables:

(i) conditions in which they have advantages,

(ii) special precautions to be observed when installing,

(iii) reasons for current ratings in I.E.E. Regulations being higher than those for equivalent rubber or p.v.c.-insulated cables, and the factors which may reduce the rating in practice. [C]

9. Give a set of architect's drawings on which have been marked the disposition and requirements for all electrical outlets in a building, outline the procedure you would follow in determining: (a) total loading of the installation, (b) maximum demand, (c) capacity of the main switchboard, sub-distribution boards and sub-main cables.

Your answer should give a clear picture of your programme and stages of procedure. [C]

10. Describe briefly what system of wiring and what type of switchgear, and fittings you would install in three only of the following buildings:

(a) a paint works where inflammable vapours are present,

(b) a paper board mill where floors are wet, and the air damp,

(c) a milking shed in a modern dairy farm,

(d) a block of flats where all work is to be concealed.

11. In supervising the installation of a system using p.v.c. cables in steel trunking what would you consider to be the essentials to ensure sound construction, long life, and compliance with the I.E.E. Regulations? [C]

12. Describe the requirements of the Electricity Supply Regulations, 1937, in respect of:

(a) entry of underground cables;

(b) leakage of current to earth;

(c) equipment placed on premises by the undertaking;

(d) protection of consumers' installation against excess energy.

[C]

13. Explain, with the aid of sketches where necessary, methods of overcoming problems likely to be encountered when installing screwed conduit wiring systems in each of the following situations:

(a) conduits cast in situ concrete 125 mm thick;

(b) conduits to be fixed above a suspended ceiling, but the latter must not take the weight of the conduit;

(c) conduits laid in a long concrete flat roof in which provision is made for expansion of the roof due to temperature changes;

(d) conduits fixed to a steel-framed building with asbestos sheet cladding.

[C]

14. Describe the causes of corrosion and the methods of minimising corrosion when installing each of the following:

(a) unarmoured bitumen-served, lead-sheathed cables installed underground;

(b) mineral-insulated metal-sheathed cables installed outdoors;

(c) sheet steel switchgear

Chapter 9

D.C. GENERATORS AND MOTORS

9.1. General Principles

Direct-current generators (sometimes termed dynamos) and d.c. motors are somewhat similar in construction. If externally driven, a d.c. motor will deliver a direct-current supply; whilst a d.c. generator is normally capable of running as a motor. However, they will run more efficiently if applied to the purpose for which the machine is designed.

The d.c. generator is based on the fundamental principle that when a conductor is made to rotate past a d.c. magnetic field, an e.m.f. will be induced in the conductor. Conversely, the d.c. motor relies on the fact that a conductor which is carrying current and is free to move across a magnetic field then it will experience a mechanical force.

The rudimentary machine is represented in Fig. 9.1. When direct current is supplied to commutator segments 1 and 3, with the polarities as shown, the machine will act as a motor and rotate in

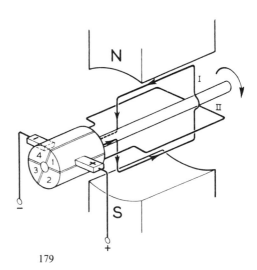

Fig. 9.1. Elementary d.c. machine which can act as either a motor or a generator.

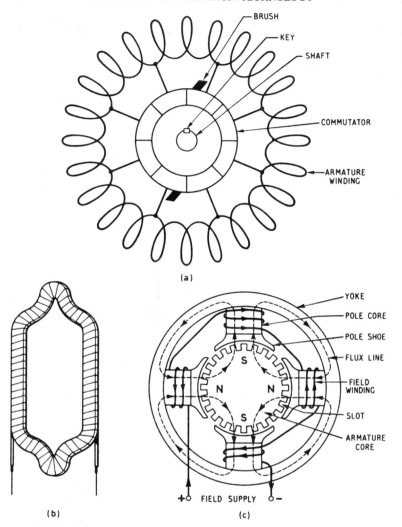

Fig. 9.2. Constructional details of d.c. machine. (*a*) Armature connexion to commutator. (*b*) Armature coil consisting of many turns. (The sides slide into slots; ends fold back.) (*c*) 4-pole field system and armature core.

a clockwise direction when looking from the commutator end. The commutator segments are insulated from the shaft and each other. The direction of motion may be found from either the left-hand or flat-hand rules. Alternatively, this direction can be determined by considering coil I as a solenoid which produces a north magnetic pole on the side nearest the observer to produce rotation by attraction to the south pole of the permanent field magnet.

As an exercise, the student should show how the 2-pole commutator machine produces continuous rotation in one direction when supplied with d.c. It is also instructive to trace out the current movement in the loops with a unidirectional supply to the brushes.

If the supply is disconnected and the shaft rotated in the same direction as before the machine will now act as a generator. From the right-hand rule, the current direction in the loop will be opposite to those as shown in the diagram but positive and negative poles at the brushes will be unchanged. Again, although the current in the loops will be alternating, direct current will be supplied to the external circuit.

In practice, many coils are required to form the *armature*, in order to produce a reasonably even d.c. supply from a generator or smooth running in a motor. The coils are connected to form a continuous closed loop, as shown diagrammatically in Fig. 9.2(*a*), tappings being made to the multi-segment commutator.

The coils of the modern drum-type armature consist of many insulated turns for normal supply voltages. For ease in manufacture they are preformed and then tightly packed in slots prior to connexion to the commutator according to the type of winding (Fig. 9.2(*b*)). The sides of the coil are one pole pitch (distance between the north or south poles) apart.

Generators and motors consist of four main parts, i.e. yoke, poles, armature, commutator and brushes. There are, however, many additional constructional features which make for increased efficiency.

9.2. Yoke

The yoke of the machine simply consists of the external frame but, as seen from Fig. 9.2(*c*), it also forms part of the field magnetic circuit. The yoke of older machines is of a heavy cast-iron construction; this has been largely replaced by a lighter fabricated mild steel casing or by cast steel having superior magnetic qualities and thus able to carry a higher flux before magnetic saturation is reached.

9.3. Poles

The poles are designed to produce the maximum flux in the slots containing the armature conductors. Each pole assembly consists of a pole core, which may be circular or rectangular, over which is threaded a field coil. The core is bolted to the yoke and has an extension consisting of the pole-shoe. The pole-shoe holds the field coil in place and produces a better flux distribution spread over the armature by means of its curved face.

The pole-shoe may be part of the pole-core and is laminated to minimise eddy currents being set up through being cut by the flux from the rotating armature conductors.

The gap between the pole-faces and the armature core is kept as small as possible, thus extreme accuracy in manufacture and assembly is essential. A small air-gap ensures a low reluctance thereby reducing the magneto-motive force (m.m.f.). This in turn reduces the exciting current and the size of the field coils needed. 'Soft' magnetic materials with high permeability also allow high flux densities before saturation is reached.

9.4. Armature

Laminations of high-quality soft-magnetic-alloy material with a low hysteresis loss are also used for the armature-core assembly. The construction is used to minimise the eddy currents which are induced through cutting the field flux on rotation. All the laminations are lightly insulated from their neighbours and tightly clamped together. Means are adopted for cooling by ventilation air ducts or holes.

Slots for housing the armature coils are parallel-sided. Wedges are often fitted to maintain these conductors in position. The coil ends (Fig. 9.2(b)), or 'overhands', are tightly pressed and tied down to prevent fouling on rotation. The coils are insulated from the metal slots by suitable paper or cloth insulation.

9.5. Commutator and Brushgear

The commutator segments are really an extension of the armature windings, the number of individual segments (Fig. 9.3) being equal to the number of slots and usually separated from each other by mica or micanite (a composition of mica and shellac).

Without brushes to make contact with the commutator, the current cannot be collected from a generator or supplied to a motor. Flat springs on the carbon or graphite brushes ensure a firm pressure. The exact brush material and pressure are selected by the manufacturers to suit the characteristics of a particular machine, although adjustment of pressure is arranged to allow for brush wear. A brush box houses each set of brushes, the brushes being connected to terminals by flexible pigtails and often tilted to produce a larger contact area.

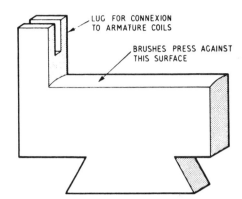

Fig. 9.3. A typical copper commutator segment.

9.6. Lap and Wave Windings

Armature coils can be connected to the commutator in various ways, although general practice has reduced them to two forms—wave winding and lap winding. For wave winding, the coils are joined to produce two parallel paths between the positive and negative brushes. With the lap winding, the total number of parallel circuits is equal to the number of poles—this may be remembered by the letter 'p' in 'laP' and the same letter in 'Poles'.

For the larger sizes of generators and motors, especially with four or more poles, the lap method is more advantageous. Since this method provides additional parallel paths, it follows that for a given machine heavier loads can be carried because the current through each armature conductor is reduced.

9.7. Armature Reaction and Interpoles

When armature conductors are rotated in a generator, the flux set up by the current in these conductors affects the distribution of the main field. The effect on a 2-pole field is considered in Fig. 9.4.

Since lines of force cannot cross, the field becomes distorted. The resultant direction of the field may be obtained by addition of the flux phasors. This general effect is known as armature reaction; in a generator, it is seen to twist the main field in the direction of rotation (the reverse occurring in a motor). Also, the main flux becomes weakened bringing a reduction in generated e.m.f. A further point to note is that the armature flux is not constant but depends upon the armature current.

The armature cross-field can be neutralised at all loads by fitting *interpoles*. These are small poles fitted between the main poles

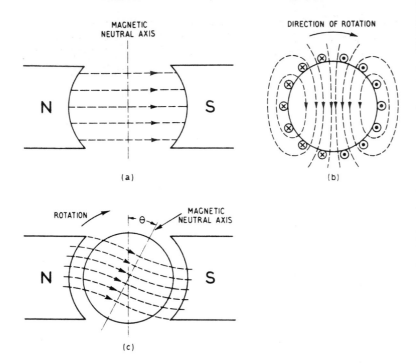

Fig. 9.4. Armature reaction caused by distortion of main field. (*a*) Flux due to field poles. (*b*) Cross flux due to armature current. (*c*) Resultant field distribution.

(Fig. 9.5), and connected in *series with the armature circuit* to provide the necessary counter-flux. The series connexion ensures proper flux correction at all loads since the interpole current and therefore field current rises and falls with that of the armature.

9.8. Commutation

Commutation is linked with the effects of armature reaction and is the term used to denote the process of producing a unidirectional, or direct, current from the alternating current generated in the armature coils as they sweep past the field pole-pieces.

In a d.c. generator, the purpose of brushes is to collect current from the armature so as to feed the external circuit. Referring to Fig. 9.4(*a*), it will be appreciated that the ideal position for these brushes is in the vertical position as shown, i.e. in the magnetic neutral axis. At this position no flux is being cut, thus there is no armature-coil e.m.f. or current and no sparking from this action should occur. However, a.c. is generated in the armature coils as they pass under alternate north and south poles. This induced e.m.f., which is proportional to the rate of change of current, sets up a current which is short-circuited as the brushes pass over the commutator segments.

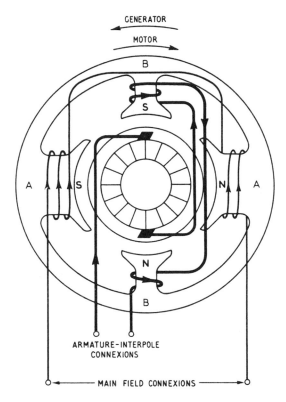

Fig. 9.5. Interpole connexions and polarity. AA—main field poles, BB—interpoles.

The use of interpoles will tend to minimise the effects of sparking produced by the induced current set up by short-circuited armature coils. Correct interpole connexion for generators or motors, as shown in Fig. 9.5, sets up a neutralising flux to that causing the back e.m.f. Compensating windings placed in slotted pole shoes and connected in series with the armature act in the same way and may alternatively be used for this purpose.

9.9. E.M.F. Equation

When an armature (generator or motor) turns through one complete revolution, each conductor of the armature coils cuts the flux leaving the north pole and entering the south poles (i.e. twice the number of poles).

Let p = pairs of poles

Φ = flux per pole in webers

N = rotational speed of armature in rev/min

Z = number of armature conductors

C = number of *parallel* paths through armature between brushes of opposite polarity

     = 2 for wave and $2p$ for lap winding

$\therefore$ $\dfrac{Z}{C}$ = number of armature conductors in series for each parallel path

In one revolution, each conductor cuts across a magnetic flux equivalent to $2p\Phi$ Wb.

$\therefore$ Total flux cut by 1 conductor/second $= \dfrac{2p\Phi N}{60}$ Wb

$\therefore$ Average e.m.f./conductor $= \dfrac{2p\Phi N}{60}$ volts

(1 volt is induced when a flux of 1 weber is cut/second)

$\therefore$ Total e.m.f., $E = \dfrac{2p\Phi N}{60} \times \dfrac{Z}{C}$ volts

As the armature rotates, an e.m.f. is generated in its coils. This happens in a motor as well as in a generator and although—as given by Lenz's law—it will be in the nature of a back e.m.f. in the motor, the calculations will be exactly the same as for the generator.

Example 9.1

Distinguish between lap-connected and wave-connected armature windings, and deduce the e.m.f. equation for a d.c. machine.

Calculate the flux in a 4-pole dynamo with 722 armature conductors generating 500 V when running at 1000 rev/min if the armature is (a) lap connected, and (b) wave connected.   [C]

Description already given.

(a) Lap $E = \dfrac{2p\Phi N}{60} \times \dfrac{Z}{C}$

$\therefore$ Flux/pole, $\quad \Phi = \dfrac{E \times 60C}{2pNZ}$

As C = number of poles = 4

$$\Phi = \frac{500 \times 60 \times 4}{2 \times 2 \times 1000 \times 722} \qquad = 0.0416 \text{ Wb}$$

(b) Wave $\qquad \Phi = \dfrac{E \times 60C}{2pNZ}$

As $C = 2$ $\qquad \Phi = \dfrac{500 \times 60 \times 2}{2 \times 2 \times 1000 \times 722} \qquad = 0.0208 \text{ Wb}$

[C]

Example 9.2

If the flux/pole in Example 9.1 were reduced to 0.01 Wb, at what speed must the armature be rotated to generate the same voltage?

Since Z, p and C are constant for a given machine, they may be represented by a single constant K. Therefore, $E = K\Phi N$

(a) Lap

$$\frac{E_1}{E_2} = \frac{K\Phi_1 N_1}{K\Phi_2 N_2} = \frac{\Phi_1 N_1}{\Phi_2 N_2}$$

where the subscripts 1 and 2 stand for the old and new values respectively

$\therefore \qquad N_2 = \dfrac{E_2 \Phi_1}{E_1 \Phi_2} \times N_1$

As $E_2 = E_1$ $\qquad N_2 = \dfrac{0.0416 \times 1000}{0.01} \qquad = 4160 \text{ rev/min}$

(b) Wave

Since flux is halved, $\qquad\qquad\qquad$ Speed = 2080 rev/min

9.10. D.C. Generators

The machines are named accordingly to the form of field-armature connexions (Fig. 9.6); field coils provide the necessary excitation.

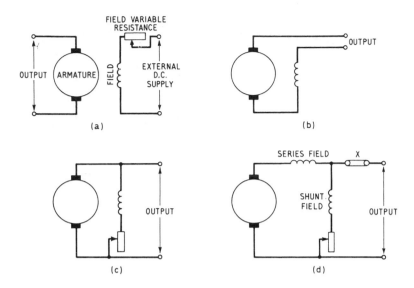

Fig. 9.6. Types of d.c. generators. (*a*) Separately-excited. (*b*) Series. (*c*) Shunt. (*d*) Short shunt compound (long shunt when series field connected at X).

9.11. Separately-excited Generators

Machine graphs or characteristic curves (often abbreviated to 'characteristics') are invaluable in understanding the working or performance of apparatus. Although there may be individual variations these curves follow a general pattern and may be obtained from tests and design information.

If at no load the separately excited generator is externally rotated at normal speed, the output or terminal voltage is proportional to the magnetic pole flux as obtained by the field current; the graph shape follows on the lines of the *B–H* curve (Fig. 9.7). This is known as the open circuit characteristic. The illustration shows that the curve does not start at the origin (O). The height above this point to the commencement of the curve represents the residual magnetism due to hysteresis.

The load characteristic (Fig. 9.7) shows a falling off of voltage as the load increases and is typical of this type of machine. Armature reaction weakens the field and is thus a further contributory factor, but the main cause of the voltage fall is due to the volt drop in the armature itself. This leads to the basic generator expression.

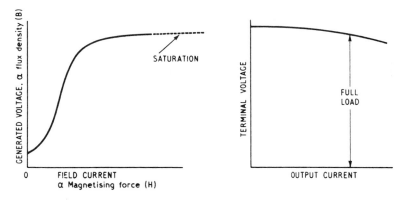

Fig. 9.7. Characteristics of separately excited generator. (*left*) Open circuit. (*right*) The generator on load.

$$V = E - I_a R_a \quad \text{where} \quad V = \text{terminal voltage}$$
$$E = \text{voltage generated}$$
$$I_a = \text{armature current}$$
$$R_a = \text{armature resistance}$$
$$\text{(including interpoles)}$$

9.12. Series Generator

Here the circuit must be completed by the load before generation can take place. As seen by the load characteristic (Fig. 9.8), the

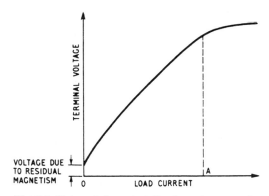

Fig. 9.8. External characteristic of a series generator (the load voltage is proportional to the current for the range O to A).

generator is totally unsuitable for normal purposes since the voltage is irregular, being proportional to the load current. The machine however comes into its own as a booster, where it is necessary to compensate for an IR voltage drop in long feeder lines.

Since the field windings are in series with the full-load current they are made up of relatively few turns which is sufficient to provide the necessary magnetising force or ampere-turns.

9.13. Shunt Generator

In contrast to the series type, the field winding being in parallel with the armature and load consists of a large number of light turns. Also the field current is not affected to any great extent by change in the load current.

The open circuit characteristic is the same as for the separately excited generator. Since an increase in field current produces a greater magnetic flux, up to saturation, and thus in turn increases the voltage generated, *voltage control can be easily effected by means of a variable resistor in the field circuit.*

Above a certain value of field resistance generation does not become possible, the resistance in ohms being known as *Critical Resistance* and can be found by the slope of the line OA (Fig. 9.9). At this particular position it is tangent to the curve at the origin. Thus the critical resistance is equal to $\dfrac{\text{generated e.m.f.}}{\text{field current}}$ corresponding to any point on OA. As the slope is reduced, the line intersects the open circuit curve and a voltage will be indicated at the terminals. OB

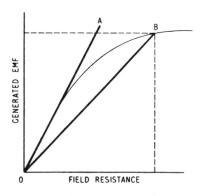
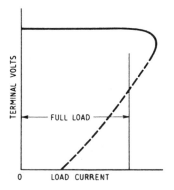

Fig. 9.9. Shunt generation characteristics. (*left*) Open circuit characteristic. OA represents critical resistance line, OB represents resistance line for supply voltage. (*right*) External characteristic.

shows where a resistance line might cut the curve for the normal supply voltage of the generator at its rated excitation.

The residual flux in the magnetic poles, although small, is of great importance, since it enables the generator to start generating as soon as the armature is rotated. This flux generates a small e.m.f., producing a small field current making for an increase in flux and further increase in e.m.f.

The load characteristic (Fig. 9.9) illustrates how the voltage falls sharply as the load current is increased beyond full load. This is due to the combined effects of armature current and armature-reaction voltage drops. The fall in terminal p.d. produces a reduction in field current which further reduces the voltage; the machine then becomes unstable as the voltage collapses to zero value.

For the normal load-current range, the voltage is reasonably steady and the d.c. shunt generator is thus extremely useful for battery charging.

9.14. Compound-wound Generators

As we have seen the load characteristic of the series generator gives a rising characteristic while the shunt type shows a drooping curve. Since a compound machine contains both a series and shunt field winding then by suitable proportions of excitation produced by these fields, control of the terminal voltage at various loads can be obtained. The machine is said to be under-, level- or over-compounded when the terminal voltage for various loads is respectively dropping, uniform or rising.

GENERATOR WORKED EXAMPLE

Example 9.3

Explain the function of a commutator in a direct-current generator.

A load of 19·2 kW is supplied from the terminals of a 2-pole shunt generator at 240 V. The shunt winding of the generator has a resistance of 96 Ω, and the resistance of the armature is 0·2 Ω. There is a brush-contact volts drop of 2 V.

Calculate (a) the current in the armature, (b) the e.m.f. generated, and (c) the copper losses in the machine. [C]

The function of a commutator has already been described (Section 9.5).

(a) Load current $I = \dfrac{P}{V} = \dfrac{19 \cdot 2 \times 1000}{240}$ = 80 A

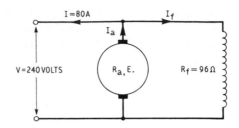

Fig. 9.10. Symbols for shunt generator calculations.

I_f = Field current. R_f = Field resistance.
I_a = Armature current. R_a = Armature resistance.
I = Load current. V = Terminal voltage.
E = Generated volts.

Inserting values in a circuit diagram (Fig. 9.10).

Since the arrangement, in effect, forms a parallel circuit and the terminal voltage is directly across the field resistance,

$$I_f = \frac{V}{R_f} = \frac{240}{96} \qquad\qquad = 2.5 \text{ A}$$

Also armature current $I_a = I + I_f = 80 + 2.5 \qquad = \underline{82.5 \text{ A}}$

(b) Terminal voltage = generated volts − armature volt drop − brush contact volts.

i.e. $\qquad\qquad V = E - I_a R_a - 2$

∴ $\qquad\qquad E = V + I_a R_a + 2$

$$= 240 + 82.5 \times 0.2 + 2 \qquad = \underline{258.5 \text{ V}}$$

(c) Total copper losses = losses in armature + losses in field,

$$= I_a^2 R_a + I_f^2 R_f$$

$$= 82.5^2 \times 0.2 + 2.5^2 \times 96$$

$$= 1361.25 + 600 \qquad\qquad = \underline{1961.25 \text{ watts}}$$

A useful exercise is to calculate the electrical *efficiency* of the above generator:

Because of the losses the actual power generated to produce the rated output

$$= 19\,200 + 1961.25 \qquad\qquad = 21.161 \text{ kW}$$

$$\text{Efficiency} = \frac{\text{output}}{\text{input}} = \frac{19.2}{21.161} \times 100 \quad = 91\,\%$$

In order to ascertain the *mechanical efficiency*, the torque of the driving motor or prime-mover would have to be given. The overall efficiency of the complete set combines both the electrical and mechanical efficiencies, numerically being obtained by their product and often termed the *commercial efficiency*.

9.15. Motor Enclosures

Details of a modern type of d.c. motor are illustrated in Fig. 9.11. Early motors were open type and provided no electrical or mechanical protection. These are no longer permitted and four main types of enclosures are recognised (see Table 10.1).

TABLE 10.1. MAIN TYPES OF MOTOR ENCLOSURES

SCREEN-PROTECTED. The covers consist of perforated sheet metal which may be easily removed for inspection purposes. The perforations permit cool air to pass through the machine but provide little protection against the entry of dust or dirt. May be used in clean dry situations.

DRIP-PROOF. To guard against the effects of dripping water the perforations are replaced by louvres of cowls which may be backed by a wire mesh to prevent small materials being drawn in by the action of the motor fan.

PIPE-VENTILATED. Where operating in very dusty or dirty conditions, the motor interior would soon become clogged, causing over-heating. With the pipe-ventilated enclosure the machine is sealed and ventilation is provided by a pipe system which brings clean air from an external clean atmosphere which may be outside the building itself.

TOTALLY-ENCLOSED. Since the dissipation of heat is only possible through the casing, this machine must be given a higher rating for the same output power than one which allows freer air circulation.

These enclosures also apply to generators and other types, such as tropic proofing, can be specified for special purposes. *Encapsulated* windings are becoming more common. The end windings and slots are made into a solid mass by epoxy-resin encapsulation, making these parts completely impervious to dust, alkalis and most acids.

9.16. Back E.M.F. and Speed Control

The cutting of the main field by the armature conductors produces a back e.m.f. in the armature; the direction of the induced e.m.f. could also be obtained by the right-hand rule. This back e.m.f. acts in opposition to the supply voltage and, for normal running conditions,

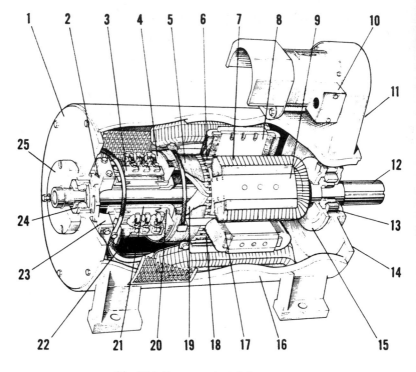

Fig. 9.11. Screen-protected d.c. motor.

1. End bracket.
2. Brush arm adjustment plates.
3. Brush arm.
4. Commutator segment.
5. Steel wire banding.
6. Armature slot wedges.
7. Field coil.
8. Interpole.
9. Mainpole.
10. Fan unit terminal box.
11. Motor driven fan unit.
12. Shaft.
13. Roller bearing.
14. End bracket.
15. Inner bearing cap.
16. Magnet frame.
17. Interpole coil.
18. Armature core.
19. Armature winding.
20. "V"-shaped end ring.
21. Commutator bush.
22. Brush holder.
23. Bearing cap/brush rocker.
24. Ball bearing.
25. Outer bearing cap.

is always a little less than the supply p.d., the difference being the armature voltage drop. From these facts the basic motor formula can be obtained.

Armature voltage drop = Applied voltage − back e.m.f.

∴ $\quad I_aR_a = V - E$ where $\quad V$ = applied voltage

or $\quad V = E + I_aR_a$ $\qquad\qquad E$ = back e.m.f.

$\qquad\qquad\qquad\qquad\qquad\quad I_a$ = armature current

$\qquad\qquad\qquad\qquad\qquad\quad R_a$ = armature resistance

As already noted

$\qquad\qquad E \propto \Phi N$ $\qquad$ where $\quad \Phi$ = flux/pole

Transposing $\qquad\qquad\qquad\qquad\qquad N$ = speed in rev/min

$$N \propto \frac{E}{\Phi}$$

The speed is thus inversely proportional to flux. From this it follows that *speed control of a d.c. motor can be simply effected by means of a variable resistor in the field circuit.* Rectification of a.c. by means of the gas-filled valve or thyratron—and latterly by the silicon-controlled rectifier—as applied to d.c. motors has brought a new means of speed control of motors from a.c. mains.

Example 9.4

Explain the function of a variable resistance in series with the shunt winding of a d.c. motor.

A 6-pole 500-V shunt motor is wave wound with 360 armature conductors. The armature resistance is 0·18 Ω and the flux/pole is 0·05 Wb. Calculate the respective speeds of the motor when the armature currents are 20 A, 60 A and 95 A. [C]

The explanation is given in Section 9.16.

$$E = V - I_aR_a \qquad (1)$$

Also

$$E = \frac{2p\Phi N}{60} \times \frac{Z}{C} \qquad (2)$$

Equating (1) and (2)

$$V - I_aR_a = \frac{2p\Phi N}{60} \times \frac{Z}{C}$$

Transposing

$$N = \frac{60C(V - I_aR_a)}{2p\Phi Z}$$

Speed for armature current of 20 A

$$= \frac{60 \times 2(500 - 20 \times 0{\cdot}18)}{2 \times 3 \times 0{\cdot}05 \times 360}$$

(Note $C = 2$ parallel paths for wave-wound armature)

$$= \frac{496{\cdot}4}{0{\cdot}9} \qquad\qquad = 552 \text{ rev/min}$$

At 60 A,   Speed $= \dfrac{500 - 60 \times 0{\cdot}18}{0{\cdot}9}$ $= 544 \text{ rev/min}$

At 95 A,   Speed $= \dfrac{500 - 95 \times 0{\cdot}18}{0{\cdot}9}$ $= 537 \text{ rev/min}$

Example 9.5

A d.c. shunt motor connected across a 240-V d.c. supply has a no-load speed of 1480 rev/min. The current input at no load is 4 A, and at full load the current input is 38 A. The shunt field windings have a resistance of 150 Ω and the resistance of the armature is 0·25 Ω. Ignore volts drop due to brushes. Calculate the speed at full load, ignoring the effects of temperature rise in the machine.   [C]

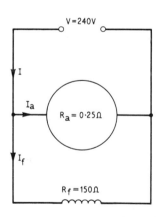

Fig. 9.12. Shunt motor problem referred to in text.

Referring to shunt-motor circuit diagram (Fig. 9.12) and with usual symbols:

$$I_f = \frac{V}{R_f} = \frac{240}{150} \qquad\qquad = 1{\cdot}6 \text{ A}$$

Load current $\qquad I = I_a + I_f$

$\therefore \qquad\qquad I_a = I - I_f$

AT NO LOAD

$$I_a = 4 - 1\cdot6 \qquad\qquad = 2\cdot4\ \text{A}$$

$$E = V - I_a R_a$$

Back e.m.f. at no load

$$E_1 = 240 - 2\cdot4 \times 0\cdot25 \qquad = 239\cdot4\ \text{V}$$

AT FULL LOAD

$$I_a = 38 - 1\cdot6 \qquad\qquad = 36\cdot4\ \text{A}$$

$$E_2 = 240 - 36\cdot4 \times 0\cdot25 \qquad = 230\cdot9\ \text{V}$$

Since the back e.m.f. is proportional to speed and calling the speeds at no-load and full-load respectively N_1 and N_2,

$$\frac{N_1}{N_2} = \frac{E_1}{E_2}$$

$$N_2 = N_1 \times \frac{E_2}{E_1} = 1480 \times \frac{230\cdot9}{239\cdot4} \qquad = \underline{1428\ \text{rev/min}}$$

9.17. Motor Power and Torque

Work is said to be done when a force is exerted against a resistance through a distance. Thus *work* is measured by *force* multiplied by *distance* moved.

Consider a motor pulley wheel on rotation (Fig. 9.13), then the work done by the tangential force in one revolution of the pulley

$$= \text{Force} \times \text{circumference}$$

$$= F \times 2\pi r$$

$$= 2\pi T \qquad \text{where } T \text{ is the torque in newton metres}$$

$$\text{Work/min} = 2\pi N T \qquad \text{where } N \text{ is the speed in rev/min}$$

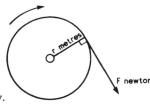

Fig. 9.13. Work done on rotation by pulley.

$\therefore$

$$\text{Power exerted} = \frac{2\pi N T}{60} \quad \text{newton metre/second (Nm/s)}$$

$$= \frac{2\pi N T}{60} \quad \text{watts, since 1 Nm} = 1 \text{ joule}$$

Example 9.6

A cutting tool exerts a tangential force of 100 N on a steel bar of 0·3 m diameter, which is being turned in a simple lathe. The lathe is driven by a chain at 800 rev/min from a 240-V d.c. motor which runs at 1800 rev/min.

Calculate the current taken by the motor if its efficiency is 85%.

What size is the motor pulley if the lathe pulley has a diameter of 0·675 m?   [C & G]

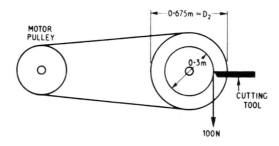

Fig. 9.14. Illustrating motor-lathe problem (see text).

$$\text{Torque} = Fr = 100 \times 0·15 \qquad = 15 \text{Nm}$$

$$\text{Power of lathe} = \frac{2\pi N T}{60} = \frac{2\pi \times 800 \times 15}{60} = 1257 \text{ W}$$

$$\text{Motor output} = \frac{1257 \times 100}{85} \qquad = 1478 \text{ W}$$

$$\text{Motor current} = \frac{\text{power}}{\text{voltage}} = \frac{1478}{240} \qquad = \underline{6·2 \text{ A}}$$

Since peripheral speeds at both pulleys are the same

$$\pi D_1 N_1 = \pi D_2 N_2 \quad \text{where } N_1, N_2 \text{ are the motor and lathe speeds respectively}$$

$\therefore$ Diameter of motor pulley $D_1 = \dfrac{D_2 N_2}{N_1} = \dfrac{0.675 \times 800}{1800} = \underline{0.3 \text{ m}}$

Example 9.7

(a) State FOUR losses which may occur in a d.c. shunt motor.

(b) A 2-kW 200-V shunt-wound d.c. motor runs at 16 rev/s and takes a current of 14 A when developing a full load output. The armature resistance is 0.4 Ω and the field resistance is 160 Ω. Calculate at full load:

(i)   armature current;
(ii) back e.m.f.;
(iii) the efficiency;
(iv) torque.   [C]

(a) I^2R losses in armature; I^2R losses in field; magnetic losses (hysteresis and eddy current; friction and windage.

(b) Field current $I_f = \dfrac{V}{R_f}$ where V = supply voltage
                               R_f = field resistance

$\qquad\qquad = \dfrac{200}{160}$ $= \underline{1.25 \text{ A}}$

(i) Supply current $I = I_a + I_f$ where I_a = armature current
                                   I_f = field current

$\qquad\qquad \therefore I_a = 14 - 1.25$ $= \underline{12.75 \text{ A}}$

(ii) $V = E + I_a R_a$ where E = back e.m.f.

$\qquad\qquad E = V - I_a R_a$
$\qquad\qquad\; = 200 - 12.75 \times 0.4$ $= \underline{194.9 \text{ V}}$

(iii) Efficiency $\quad = \dfrac{\text{Output}}{\text{Input}}$

$\qquad\qquad\quad = \dfrac{2000}{200 \times 14} \times 100$ $= \underline{71.4\%}$

(iv) Power, $\qquad P = 2\pi nT$ where n = rev/s
                                    T = torque

$\qquad\qquad T = \dfrac{P}{2\pi n}$

$\qquad\qquad\; = \dfrac{2000}{2\pi \times 16}$ $= \underline{19.9 \text{ N m}}$

The basic equation for the d.c. motor, as already noted, is

$$V = E + I_a R_a \quad \text{where } V = \text{applied voltage}$$
$$E = \text{back e.m.f.}$$
$$I_a R_a = \text{armature voltage drop}$$

Multiplying through by the armature current,

$$V I_a = E I_a + I_a{}^2 R_a$$

Thus the total power supplied ($V I_a$) is equal to the power developed by the motor ($E I_a$) plus the armature heat losses ($I_a{}^2 R_a$).

We can now equate equivalent electrical and mechanical powers:

i.e. $\qquad E I_a = \dfrac{2\pi N T}{60}$ $\qquad$ where p = pair of poles

$\qquad\qquad\qquad\qquad\qquad\qquad\qquad \Phi$ = flux/pole

but $\qquad E = \dfrac{2\pi N p}{60} \times \dfrac{Z}{C}$ $\qquad Z$ = no. of armature conductors

$\qquad\qquad\qquad\qquad\qquad\qquad\qquad C$ = no. of parallel paths

$\therefore \qquad I_a \times \dfrac{2\pi N p}{60} \times \dfrac{Z}{C} = \dfrac{2\pi N T}{60}$

$\therefore \qquad\qquad\qquad T = 0\cdot 318 I_a \Phi p \times \dfrac{Z}{C} \text{ newton metres}$

$\therefore$ $\quad T \propto I_a \Phi$ since the other values may be taken as constant, *from which it follows that the total torque developed by a d.c. motor is proportional to the product of the armature current and the flux per pole.* The useful torque is less than this since account must be taken of the losses.

9.18. Series Motor

As in the generator, the field also carries the main current and the winding must consist of a number of heavy turns. The torque is proportional to current times flux ($I\Phi$). Since the flux is proportional to the current until magnetic saturation is reached

$\therefore \qquad\qquad\qquad\qquad T \propto I^2$

thus double the load current increases the torque four times.

From the series characteristic (Fig. 9.15) it may be seen that at low speeds there is a heavy input current and torque. Consequently, the

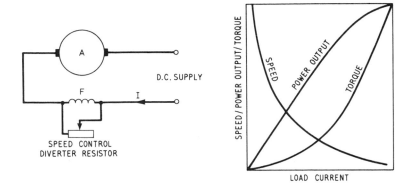

Fig. 9.15. Series motor connexions and characteristics.

series motor may start against load. Conversely, 'racing' occurs at no-load so that a positive drive is advocated, rather than by a belt which may snap at any time. The sphere of application is lift motors, cranes, trains and other forms of traction where heavy starting duties may arise.

For any given load, speed control from no-load to full-load may be effected by means of a series resistor since the speed of the motor decreases with terminal voltage. Because this resistor would have to carry a heavy current, a diverter in the form of a variable shunt winding is often employed.

The direction of rotation may be reversed by either interchanging the armature *or* the field terminal leads.

9.19. Shunt Motor

Since the field winding is across the mains, the flux and back e.m.f. are almost constant under normal conditions. Therefore, this motor is considered as a constant speed type. With increase in load there will be some drop in speed (Fig. 9.16) due to the field distortion and subsequent weakening caused by armature reaction. However, this drop in speed may be offset by the weakening of the flux. Speed control, as explained in Section 9.16, may be effected by a variable resistor in the shunt circuit, although speed increase is limited to about three times full-load speed.

Speed reduction only is possible by placing a variable resistor in series with the armature and is considered more wasteful than the previous method, as the resistor has to carry the full armature current.

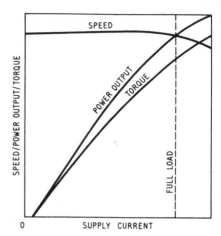

Fig. 9.16. Shunt motor
characteristics.

The characteristic therefore depends upon the ratio of shunt and series ampere-turns. If the shunt coil is wound to produce a weak flux, the machine will almost run as a series motor and vice versa.

The motor has the disadvantage of a low starting torque and it is therefore advisable to start with the load disconnected. The torque increases with load current.

The shunt type is the most common of all d.c. motors because of its constant-speed characteristic. Its ease of speed variation for any particular load also makes it a rival to a.c. motors where speed control is considered necessary.

Similar to the series motor, the direction of rotation may be reversed by interchanging either the armature-terminal or the field-terminal leads. The brush positions may require adjustment to suit rotation.

9.20. Compound-wound Motor

The flux produced by each of the field poles is the resultant of the flux produced by the shunt and series coils; it also depends upon the direction of these field windings, which may be connected as

CUMULATIVE where the fluxes from the shunt and series coils assist each other, or

DIFFERENTIAL where they are arranged in opposition so that the resultant flux is weakened, leading to a rise in speed on load.

The characteristic therefore depends upon the ratio of shunt and series ampere-turns. If the shunt coil is wound to produce a weak flux, the machine will almost run as a series motor and vice versa.

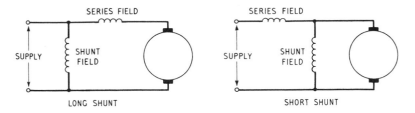

Fig. 9.17. Compound-wound motor connexions.

Thus, the compound motor (Figure 9.17) may be designed to possess the best of both worlds, that is a good starting torque together with an almost constant speed at all loads. The application is wide, being limited only by cost; a typical use is for rolling mills which are subjected to heavy fluctuations in load.

STARTING. A resistor must be connected in series with the armature as starting takes place. This is because the supply as it passes through the low resistance of the armature would otherwise produce a damaging current. As the motor picks up speed the back e.m.f., which is induced, has the effect of maintaining the current at a safe value while the starting resistance is being gradually cut out. Figure 9.18 depicts the type as used for a shunt motor.

The no-volt trip, which is a form of electromagnet, has a number of functions. It maintains the starter handle at the ON position when normal full speed is reached. Should the supply fail or a break occur

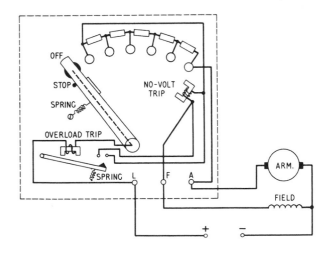

Fig. 9.18. D.C. faceplate starter.

in the field circuit, the trip operates and the starter handle springs back to the OFF position.

The overload trip, which is wound with relatively few turns of heavy wire and is usually adjustable, operates when excess current occurs. This trip short-circuits the no-volt device so that it loses its magnetic holding power under these circumstances.

For series motors the variable resistor used for speed control may also be employed for starting purposes.

9.21 Automatic Starting of D.C. Motors

When d.c. motors are to be started by push-button, the switching of the starting resistors is controlled automatically, usually by means of electromagnetic contactors and timing relays. In one form of control a series of graduated time-delay contactors are arranged so that the added resistance in the armature circuit is gradually cut out as the motor picks up speed and generates an increasing back e.m.f.

An ingenious type is known as a *counter-e.m.f. starter*. Referring to Fig. 9.19, the start button operates contactor C_1 which puts the starting resistance in series with the armature. The armature counter- or back-e.m.f. is proportional to the motor speed. As the latter increases, the back e.m.f. which is across the operating coil of contactor C_2 increases until it is sufficient to close the contactor, thereby short-circuiting the starting resistance.

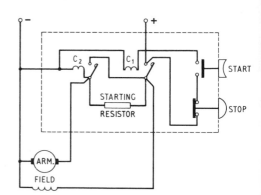

Fig. 9.19. Counter e.m.f. starter.

These starters can operate equipments that are switched automatically such as by the float switch of pumping equipment or by the pressure switch of an air compressor.

Example 9.8

Calculate the ohmic value of a starting resistance for the following d.c. shunt motor: output, 14 920 W; supply, 240 V; armature resistance, 0·25 Ω; efficiency at full load, 86 per cent.
The starting current is to be limited to 1½ times full-load current. Ignore the current in the shunt windings.   [C]

$$\text{Efficiency} = \frac{\text{output}}{\text{input}}$$

$$\therefore \qquad \frac{86}{100} = \frac{14\,920}{240I}$$

$$\text{Full-load current } I = \frac{14\,920 \times 100}{240 \times 86} \qquad = 72\cdot3 \text{ A}$$

$$\text{Starting current} = 72\cdot3 \times 1\tfrac{1}{2} \qquad = 108\cdot45 \text{ A}$$

$$\text{Total resistance} = \frac{V}{I} = \frac{240}{108\cdot45} \qquad = 2\cdot213 \ \Omega$$

$$\therefore \qquad \text{Starting resistance} = 2\cdot213 - 0\cdot25 \qquad = \underline{1\cdot963 \ \Omega}$$

9.22. Ward Leonard Speed Control

The exciter, driving motor and generator are mechanically coupled (Fig. 9.20). The purpose of the exciter, which is in effect a miniature generator, is to supply a constant d.c. voltage to the generator and motor fields, although with d.c. mains available the exciter would not be necessary.

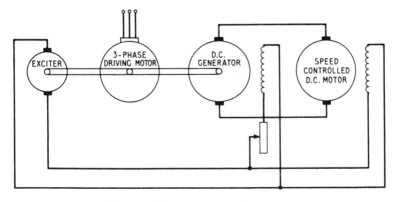

Fig. 9.20. The Ward Leonard system.

Speed control is effected by the variable resistor in the generator field circuit. This varied excitation controls the voltage to the d.c. motor thereby varying its speed, which can be changed smoothly from zero to full load speed.

In practice, the variable resistor could be replaced by a potentiometer which also allows for reversal of the motor field current and can thereby permit the machine to rotate in a reverse direction.

EXERCISES

1. (*i*) Explain with the aid of sketches, how 'commutation' is effected in a d.c. generator.

(*ii*) Explain briefly the effects of 'armature reaction' in this generator. [C]

2. Describe and explain, with diagrams, methods of reversing the direction of rotation of (*a*) a d.c. series motor, (*b*) a d.c. shunt motor. A 4-pole, 500 V, d.c. shunt motor is lap wound with 640 conductors. The armature resistance is 0·22 Ω, and the flux per pole is 0·04 Wb.

Calculate the respective motor speeds when the armature currents are 10 A, 50 A, and 90 A. [C]

3. Explain with the aid of diagrams, how speed control of the following may be obtained: (*a*) d.c. shunt motor, (*b*) three-phase induction motor with wound rotor. In each case give reasons why speed control is obtained by the method used. [C]

4. For what purpose is a variable resistance sometimes connected to the shunt circuit of a d.c. generator?

A d.c. generator with a terminal voltage 240 V, supplies a load of 18·84 kW. The armature resistance is 0·15 Ω, and there is a total brush-contact drop of 2 V. The resistance of the field windings is 108 Ω.

Calculate (*a*) the armature current, (*b*) the generated e.m.f., (*c*) the total copper losses in the windings.

5. Give the I.E.E. Regulations for (*a*) Control of motors, (*b*) Motor circuits. [C]

6. Explain, with diagrams, two methods of varying the speed of a d.c. shunt motor. Give an indication of the range of speed control attainable, and the advantages and disadvantages of each method.

A d.c. shunt motor connected across a 240-V d.c. supply has a no-load speed of 1480 rev/min. The current input at no-load is 4 A, and at full-load the current input is 38 A. The shunt field windings have a resistance of 150 Ω and the resistance of the armature is 0·25 Ω. Ignore volts drop due to brushes. Calculate the speed at full load, ignoring the effects of temperature rise in the machine. [C]

7. (*a*) Draw typical torque characteristics of: (i) a d.c. shunt-wound motor; (ii) a d.c. series-wound motor. Explain the shape of the curve in each case.

(b) A 460-V d.c. shunt motor running on no-load at 24 rev/s takes a current of 5 A. The resistance of the field winding is 230 Ω and the resistance of the armature is 0·2Ω. Calculate the speed of the motor when it runs with a full load input of 32 A. Assume that the field current remains constant.

[C]

Chapter 10

A.C. MOTORS

10.1. The 3-phase Induction Motor—Constructional Features

The 3-phase induction machine is very robust and, because of its simplicity and trouble-free features, is the type of motor most commonly employed for industrial use. There are only three essential parts, namely the frame, the stator winding and the squirrel-cage rotor, although there are many ancilliary features.

The stator frame consists of cast iron or, in the smaller sizes, of rolled steel and does not serve any magnetic purpose as does the yoke in d.c. machines. Its purpose is simply to supply mechanical protection and to provide support for the stator stampings, windings and arrangements for ventilation. Three symmetrically-placed windings (one for each phase) are fitted to slotted laminations which are made from high-grade alloy steel. The stator windings may be connected in star or delta; for star-delta starting the six ends of the three windings are brought out to the terminal box. Further protection may be given to the windings by a relatively new method termed *encapsulation*. It safeguards against dust, dampness, acid and alkaline atmosphere in heavy concentrations, the windings being encased in a block of cast epoxy resin, which also completely fills all spaces within the slots and coils.

For the larger sizes, the rotor winding is made up of a cylindrical frame of copper bars joined together at the two ends by brass or copper rings. Hence the name of *squirrel-cage* (s.c.) the cage forming a permanently short-circuited winding accommodated in the laminated rotor core. For motors up to 2000 W, s.c. rotors are made of high pressure aluminium, cast solid with the annealed steel-alloy laminations. The projecting blades of the cage help to provide cooling and ventilation. Motor performance is improved by the slewed rotor slots. The slewing prevents 'cogging' which may inhibit starting at a particular rotor position.

10.2. Principle of Action

The turning effort or torque produced by all electric motors depends upon the interaction of magnetic fields. The induction-type

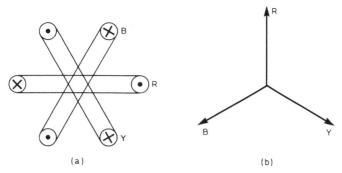

Fig. 10.1. (*a*) 3 stator windings with each phase set at 120°. (*b*) Positive directions of flux phasors for each phase, at a phase difference of 120°.

motor relies upon the production of a rotating field. In essence, this is similar to a rotating field set up by a bar magnet that is pivoted at the centre and rotated. Here, the rotating field is set up by stationary coils accommodated around the stator core.

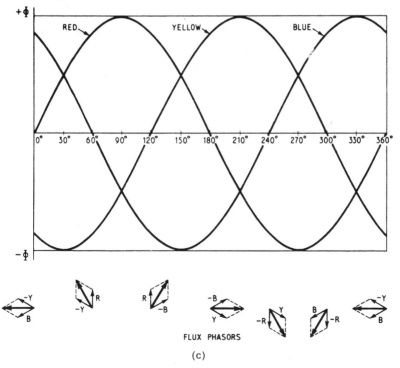

FLUX PHASORS

(c)

Fig. 10.1. (*c*) Rotating field set up by a 3-phase supply.

The 3-phase stator windings may be simplified by regarding them as three coils set at 120 degrees (Fig. 10.1(a)), corresponding flux phasors showing positive directions being depicted in Fig. 10.1(b).

Students will find it a useful exercise to trace out the creation of the rotating field. Referring to the 3-phase flux sine waves in Fig. 10.1(c), flux phasors may be taken at 60 degree intervals, the magnitude of the phasors being proportional to the ordinates in each case and the direction being taken from Fig. 10.1(b). It must be noted that the direction of each phasor is reversed when the sine flux wave is negative. The resultant of the fluxes at these 60 degree intervals is clearly seen to produce a rotating field.

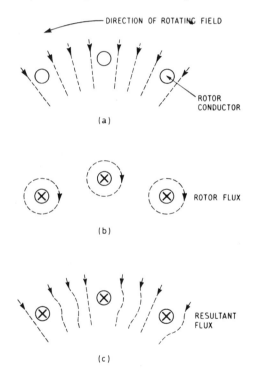

Fig. 10.2. Effect of rotating field on squirrel-cage rotor.

The rotation of the shaft to which the rotor is attached can now be seen by studying a portion of the air gap (Fig. 10.2). As the rotating field sweeps past the squirrel-cage conductors (a) it induces an e.m.f. in each conductor; the resulting current thereby producing a concentric field round each of the conductors (*b*). The direction of the induced current may be obtained by the right-hand or flat-hand rules (Section 1.7), assuming an anti-clockwise rotating field. The anti-clockwise rotation is equivalent to a movement of the rotor from right to left with the field as stationary. As lines of force do not cross, the resultant field shows a strengthening of the flux on the right-hand side of each rotor conductor and a weakening on the left-hand side. Since the conductors move from a strong field to a weak one, the rotor moves in an anti-clockwise direction (*c*), i.e. *the rotor follows the same direction as the rotating field.*

10.3. Synchronous Speed and Slip

Because the rotor follows the rotating field round, the speed of the rotating field is an important factor in determining the rotational speed of the rotor.

The speed of the rotating field depends on (1) the frequency of the supply and (2) the number of poles in the motor, according to the formula speed N (in rev/min) $= f \times 60/p$, where f = supply frequency and p = number of pairs of poles. Thus, if the 2-pole motor is operated on a supply frequency of 50 Hz, the speed of its rotating field is $50 \times 60/1 = 3000$ rev/min. The speed of the rotating field is known as the *synchronous speed* of the motor.

Clearly the rotor cannot move at this speed otherwise it would not be cut by the field. The difference between synchronous speed and the actual rotor speed is known as the slip speed, or simply as slip.

$$\text{Slip} = \text{Synchronous speed} - \text{actual speed of the rotor}$$

$$\text{Percentage slip} = \frac{\text{Synchronous speed} - \text{actual speed}}{\text{Synchronous speed}} \times 100$$

Example 10.1

A 4-pole 3-phase motor runs at 1450 rev/min. If the supply frequency is 50 Hz, what is the percentage slip?

$$\text{Supply frequency } f = \frac{pN}{60}$$

$$\text{Synchronous speed } N = \frac{60 \times 50}{2} \qquad = 1500 \text{ rev/min}$$

$$\text{Percentage slip} = \frac{1500 - 1450}{1500} \times 100 \quad = 3\tfrac{1}{3}\%$$

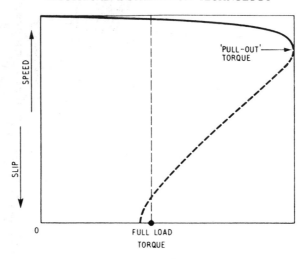

Fig. 10.3. Squirrel cage speed–torque characteristic.

10.4. Performance

The slip is small on no-load, in other words the motor runs at almost synchronous speed. As the load increases the speed falls and slip increases. The squirrel-cage machine may be considered as possessing a 'shunt' characteristic; it has a heavy starting current— about 6 times the full-load current— and poor starting torque. Figure 10.3 shows the speed–torque curve. Beyond a certain torque, called the 'pull-out' torque, the motor speed rapidly falls to zero. Whilst it varies with different makes, the maximum torque can be taken as $2\frac{1}{2}$ times that of full-load. In order to improve performance various means have been adopted. Star-delta and auto-transformer starting, as we shall see, have the effect of reducing the initial rush of current.

The squirrel-cage rotor may be replaced by the wound type, armature windings being placed in slots in the normal manner. This has the advantage that the employment of slip rings enables an external resistance to be added to that of the rotor for greater motor efficiency, and also has the effect of improving the starting torque. An ingenious scheme for the improvement of the induction motor performance, without essentially affecting the simplicity of design, is provided by the dual- or double-cage rotor. This type is sometimes termed a high-torque motor, as the starting torque is greatly improved; the starting current is also reduced.

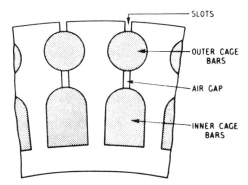

Fig. 10.4. Double cage rotor.

10.5. Double-cage Motors

An inherent feature of the induction motor is that an increase in the rotor resistance results in a higher starting torque. Here it is achieved simply by the double cage (Fig. 10.4), which takes advantage of the changes in slip frequency from 100% at standstill to a very low value at full load. The inner cage consists of bars of high-conductivity copper in deep slots. The conductors have a low resistance and a high reactance at starting and at low rotor speeds, that is, at high slip frequencies. The outer cage is placed in shallow slots and is made of bronze or similar metal so as to have a high resistance and low reactance.

When the supply is switched on there are thus small current values in the inner cage and larger currents in the outer cage. But the high resistance of the outer cage bars limits these higher currents and also results in a high starting torque. When running, there are now high currents in the inner cage (since slip frequencies are reduced) and very little current in the outer bars so that under load conditions the motor runs almost as a single-cage type.

10.6. Starting Methods

(a) DIRECT-ON-LINE (D.O.L.) STARTER. This may be used up to 3·75 kW. The essentials of the starter are given in Fig. 10.5. The starter includes a 4-pole contactor, one pole of which serves for holding-on purposes. The contactor coil acts as a no-volt release should the supply fail or an open-circuit occur.

With the isolator in the ON position and the contacts LM bridged by the starter button, the 415 V contactor operates and the supply is brought to the motor. Opening the contacts by the stop button clearly interrupts the hold-on circuit.

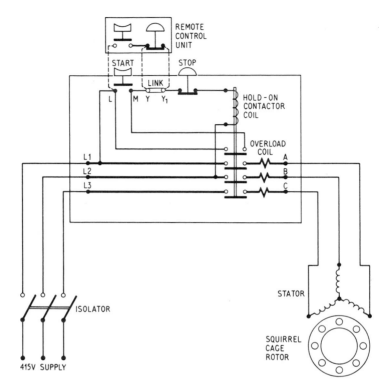

Fig. 10.5. Direct-on-line starter.

The overload coils or solenoids in some cases actuate a simple mechanism (not shown for clarity) which trips the starter at a predetermined overload current value. Adjustable oil dashpots may be fitted to prevent tripping by starting current surges. As an alternative to magnetic operation thermal overload trips based on the bimetal principle are employed especially in the smaller sizes.

It is of interest to note that extra start buttons are wired in parallel off contacts LM, while additional stop buttons are connected in series with the link YY1 removed. Only three wires are necessary for a remote combined start-and-stop button unit.

(*b*) STAR-DELTA STARTER. The six ends of the stator windings are brought out so that they can be connected in star or delta as shown in Fig. 10.6.

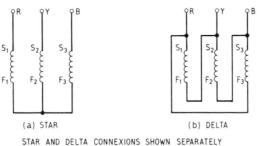

(a) STAR   (b) DELTA

STAR AND DELTA CONNEXIONS SHOWN SEPARATELY
(S-START, F-FINISH OF EACH WINDING)

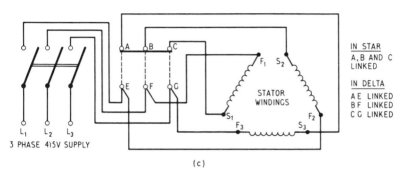

(c)

Fig. 10.6. Star-delta starter connexions.

It is important to note that these start and finish connexions must be in the correct sequence, as a reversal would result in a complete magnetic field unbalance.

Above about 2 kW the heavy starting current of squirrel-cage induction motors may produce heavy disturbances in the mains. The star-delta type of starter, which is extensively used, reduces starting current to about twice full-load current. It operates on the principle of voltage reduction. The phases are connected in star for starting and delta for running. The voltage across each phase at starting is reduced to $1/\sqrt{3}$, or 0.58, of the line volts. The current in each *phase* is reduced in the same ratio as the voltage. Therefore the line current at starting is only $1/\sqrt{3} \times 1/\sqrt{3}$ or $\frac{1}{3}$ of the current which the motor would take if switched straight on to the supply.

Unfortunately the poor starting torque, which is a feature of this machine, is even further reduced so that it cannot start against a heavy load. The double-cage rotor would effect an improvement in this respect.

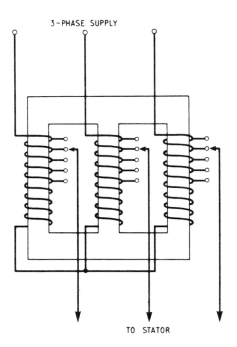

Fig. 10.7. Auto-transformer. The three variable tappings are mechanically coupled to operate together in tandem.

The modern tendency is to employ push-button starters of the automatic type. On 'start' the star connexion comes into operation by a magnetically-operated relay.

After a short interval of 1–10 seconds a thermally-operated contact or similar device causes a change to the 'run' or delta position.

(c) AUTO-TRANSFORMER STARTER OR COMPENSATOR. Sometimes voltage reduction is applied to the stator. In certain cases this is carried out by variable resistors but there are bound to be wasteful I^2R heat losses. In Fig. 10.7 the auto-transformer method is shown. The transformer core is wound on three limbs forming a closed magnetic circuit. Taps may be taken from 40 to 80 per cent of the line voltage. At a 50 % tap the current is only $(\frac{1}{2})^2$, i.e. $\frac{1}{4}$, of the full-load current. After the motor has accelerated, the stator windings are placed direct across the mains. Again, there is an automatic type where the duration of the starting periods is controlled by a timer.

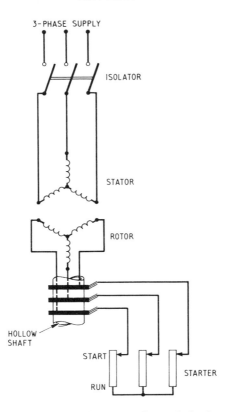

Fig. 10.8. Three-phase wound rotor induction motor.

10.7. Slip-ring Induction Motor

Reduced starting current and improved starting torque can be obtained by the insertion of added resistance into the rotor circuit through slip rings. The motor is however bound to be more expensive than the cage type since it has a 3-phase winding in the slotted rotor. Brushes bearing on the slip rings connect to a variable 3-phase resistor (Fig. 10.8). When the motor has picked up to a full speed at the 'run' position the slip rings are often short circuited. This may be achieved either manually or automatically but the short-circuiting device must be removed when the motor starts again. By completely cutting out the rotor resistance and connecting wires the losses are reduced. Efficiency may be further improved and wear eliminated by lifting the brushes at full speed.

Example 10.2

Explain the difference between a wound-rotor three-phase induction motor and a squirrel-cage induction motor. A 22·38 kW 400-V 3-phase induction motor working at full load at 0·7 power factor has an efficiency of 88 per cent. It is supplied from a motor generator consisting of a 460-V d.c. motor driving a 400 V 3-phase alternator. The respective efficiencies of the d.c. motor and of the alternator are 85 per cent and 90 per cent.

Calculate:

(a) the line current of the induction motor,
(b) the current to the d.c. motor.   [C]

See text for explanation.

(a)

$$\text{Efficiency} = \frac{\text{output}}{\text{input}}$$

Referring to induction motor

$$\frac{88}{100} = \frac{22\,380}{\sqrt{3} \times 400 \times I \times 0\cdot7}$$

∴   Motor current

$$I = \frac{22\,380 \times 100}{\sqrt{3} \times 88 \times 400 \times 0\cdot7} \qquad = \underline{52\cdot44\,\text{A}}$$

(b) Power output of d.c. motor

$$= 22\,380 \times \frac{100}{88} \times \frac{100}{90} = 28\cdot26\,\text{kW}$$

Hence

$$\text{Input to d.c. motor} = \frac{28\,260}{460} \times \frac{100}{85} \qquad = \underline{72\cdot27\,\text{A}}$$

REVERSING DIRECTION. This is easily achieved in any 3-phase induction motor by interchanging any two main or stator leads. The change in connexions has the effect of reversing the direction of the rotating field.

10.8. A.C. 3-phase Commutator Motor—Speed Control

To obtain variable motor speed on 3-phase supply an a.c. commutator motor can be used. The rotor-fed commutator motor, commonly known as the Schrage motor, provides a smooth variation

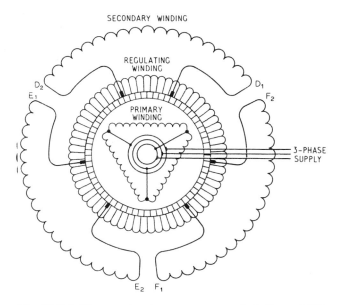

Fig. 10.9(*a*). Diagram showing arrangement of windings of AEI Schrage motor.

of speed with high efficiency, improved power factor and good starting torque. Unfortunately these advantages can only be achieved by an involved arrangement of windings and two sets of three brushes (Fig. 10.9(*a*)).

The primary winding, which sets up the rotating field, is located on the rotor, and is connected to the supply by means of slip-rings and brushes, while the secondary winding is located on the stator. This is the reverse arrangement to the ordinary induction motor. In addition to the primary winding a regulating or second rotor winding is placed in the rotor slots similar to the armature winding of a d.c. machine.

E.m.f.s are induced in the secondary winding on the stator at slip frequency; if these e.m.f.s are injected into the regulating winding at the correct phase then the speed of the motor will be increased above the normal speed. The value of the injected voltage can be controlled by a handwheel which moves the two brushes of each of the three pairs (one pair per phase) away or towards each other (Fig. 10.9(*b*)). When the brushes of each pair are on the same segment of the commutator the machine runs as a normal induction motor. A wide speed range, above and below synchronous speed, can be obtained by rotating the brushgear in one direction or the other.

The motor has a 'shunt characteristic' so that it runs at a constant torque for the normal range of load.

Starting may be quite simply achieved by means of an isolator, the brushgear being set in the minimum speed position and then the primary switch is closed.

From the basic a.c. equation $f = pN/60$, transposition gives the speed $N = 60f/p$. Thus, by interposing a frequency changer between the supply and an induction motor an alternative means of speed variation using the simple squirrel-cage induction motor is obtained. The woodworking industry, where high speeds are required for boring timber, takes advantage of this feature.

The transposed formula indicates yet another efficient means of obtaining speed control. In the *pole-changing* induction motor, the stator is wired to permit a variation in the number of poles by an external switching device. Speed changes made are in definite steps, which may be easily calculated from the basic frequency equation, according to the number of poles which are switched in or out. Halving the poles doubles the speed so that two separate stator windings may provide four distinct speeds.

10.9. Synchronous Motor

This is essentially a fixed-speed motor, the speed being independent of the load. The machine is similar to the a.c. generator and requires

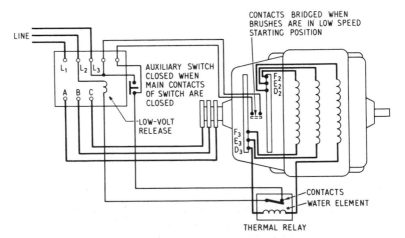

Fig. 10.9(*b*). Diagram of connexions with circuit-breaker. Thermal relay connected in secondary circuit. Mechanically-operated brushgear.

a d.c. supply via slip rings for the field excitation. One of its great advantages is that adjustment of this field leads to power-factor control—when over-excited it will take a leading power factor. Unless the rotor windings are so arranged that the machine starts as an induction motor, the synchronous motor is not self-starting but has to be run up to speed by external means.

ACTION. In the 3-phase type, the stator produces a rotating field at synchronous speed. The d.c. field supply is fed into the motor via the slip-rings, producing definite north and south poles. When the rotor is externally turned at synchronous speed, the d.c. field locks with the rotating field so that the machine continues to run at this speed.

To avoid the complication of using an external motor in order to run the synchronous motor up to speed, various types of synchronous motor that start as induction motors are available. As speed is gained the rotor is pulled into step by the d.c. field and the motor now runs as a synchronous machine. The d.c. supply is obtained from an exciter (virtually a small d.c. generator) which is often fitted to an extension of the shaft. A modern method is to obtain the d.c. supply by means of electronic rectifiers.

The synchronous-induction motor has a wound rotor and slip rings resembling the slip-ring induction motor. Salient-pole types of synchronous motor are arranged for self-starting by having, in the rotor-pole faces, either squirrel-cage windings, or a 3-phase winding which is resistance-controlled through slip rings.

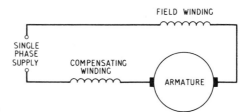

Fig. 10.10. A.C. series motor.

10.10. Single-phase Motors

Such motors cannot usually compete with the performance or efficiency of the 3-phase counterpart. They offer however a wide application in the smaller sizes, particularly for use in domestic and commercial spheres.

10.11. Single-phase Series Motor

The general arrangement may appear to be similar to the d.c. type with the addition of a compensating or neutralising winding (Fig. 10.10). We have already seen that if the main leads of a d.c.-fed series motor are reversed, the motor will continue to run in the same direction (change in direction of rotation is obtained by reversing *either* the field *or* the armature leads). Therefore the motor would run off a.c. but, due to the a.c. cycle causing inductive effects in the armature, severe sparking at the brushes is likely to occur. Further losses may take place through eddy currents in the field poles and yoke giving rise to dangerous overheating.

These difficulties may be minimised by suitable construction. One of the methods adopted to reduce sparking is the use of German silver or suitable alloy to act as high-resistance connexions to the commutator. The purpose of the compensating winding is to neutralise the inductive effects in the armature, similar in many ways to interpoles on d.c. motors. In order to reduce eddy currents the whole of the salient-field poles, where fitted, and yoke are laminated. The field winding is often distributed over the core in slots like the stator winding of a single-phase induction motor.

The motor has a 'series characteristic' so that there is good starting torque but the motor tends to race at no-load—friction at the bearings often imposes sufficient load to prevent damage. This feature which is brought out by the familiar 'whirring' sound of an electric drill, as driven by an a.c. series motor, when taken off the drilling operation. The domestic vacuum cleaner is also powered by this type of motor.

Further uses are for small lifts, cranes and compressors; its high starting torque makes the motor particularly applicable for traction purposes.

Because the machine will run on both d.c. or a.c. the series motor is often referred to as a universal motor, although the speed would be less on a.c.

10.12. Single-phase Induction Motor

The principle of action is similar to the 3-phase motor except that the rotating field must be produced from the single-phase supply. Here the rotating field is obtained by one type of split-phase circuit (Fig. 10.11). The stator has two windings—main and auxiliary—the latter being of lighter gauge although with the same number of turns and placed in slots approximately 90 electrical degrees from the main winding. The reactor helps to give 90° phase difference between the two windings. The resistance in the main winding limits the

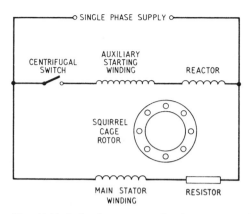

Fig. 10.11. Induction motor split-phase circuit.

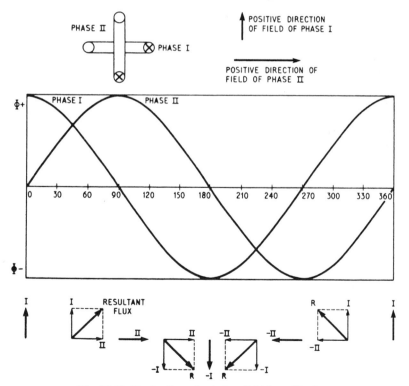

Fig. 10.12. Production of rotating field by split-phase.

starting current and is cut out once the motor reaches normal speed. Thus the motor virtually starts as a 2-phase type which produces the rotating field as may be seen from Fig. 10.12 and should be compared with the 3-phase rotating field of Fig. 10.1(c). Once speed has been reached the auxiliary winding is automatically cut out by means of the centrifugal switch. The heavy starting current may

Fig. 10.13. Single-phase capacitor motor.

be limited by the resistor in series with the main or running winding and the resistor itself may be cut out after starting has been achieved.

Characteristic of the motor is its shunt performance. The starting torque is poor, and therefore requires a loose pulley or light belt drive when the starting torque is too low, although the speed is almost constant. Removal of the starting resistance and replacement of the reactor by a capacitor still gives the dual-phase starting and produces a better starting torque. Improvement of power factor is also effected so that the capacitor-type (Fig. 10.13) is now almost universally employed for the single-phase induction motor. For economy the capacitor is of the electrolytic type and is normally strapped on to the motor.

Capacitor motors may be obtained with alternative arrangements.

1. PLAIN CAPACITOR. For smaller sizes the capacitor and starter winding may be left connected in during the whole of the running time.

2. CAPACITOR START. Centrifugal switch cuts out starting winding as already described.

3. CAPACITOR START AND RUN. Two capacitors connected in parallel are employed for starting, one of the capacitors being left in the permanently running position so as to improve the power factor.

10.13. Repulsion Motor

In appearance the repulsion motor is similar to the series commutator motor, the basic difference being that the brushes are short-circuited by a heavy conductor. It does not operate by the rotating-field effect or by the conductivity of the series motor. Rotation is produced by the reaction between magnetic fluxes created by current in the stator and the current induced in the rotor-armature by transformer action.

When the brush axis is in line with the axis of the stator winding (Fig. 10.14(a)) the current produces a polarity which, at any instant, is directly opposed to that produced in the stator. If the brushes are

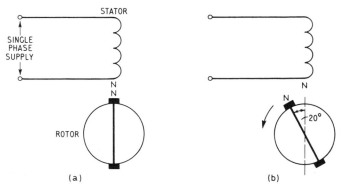

Fig. 10.14. Repulsion motor. (a) Brush axis corresponds to stator axis —no rotation. (b) Repulsion occurs when the axes are not in line.

moved to any position round the commutator except 90°, repulsion between the similar poles occur and there is continuous rotation in the direction indicated (Fig. 10.14(b)). In practice the best position is found at 20°.

The repulsion motor produces a very good starting torque. Speed control is effected by alteration in the brush position. Once again, there are a number of variations, some of which are:

1. *Repulsion-induction.* An additional squirrel-cage is placed underneath the armature winding, the main purpose of the squirrel-cage being to maintain a fairly constant speed up to full load.

2. *Repulsion-start induction motor.* The plain repulsion motor is fitted with a centrifugal device which opens a short-circuit link in the rotor when the motor reaches speed. This hybrid machine then runs as an induction motor.

10.14. Shaded-pole Motor

This is a simple type of self-starting induction motor. The motor is limited to small sizes, has a poor starting torque, and is highly inefficient, yet the ease in starting makes up for these disadvantages where loads are very light.

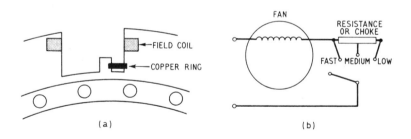

Fig. 10.15. (a) Shaded pole. Field shifts from shaded to unshaded portion. (b) Speed control of shaded pole motor.

The action is an interesting example of electromagnetism. The construction consists of a squirrel-cage motor and the stator is made up of salient poles on the lines of a d.c. machine. Each pole piece is slotted to one side and a copper ring is fitted on the smaller 'shaded' part of Fig. 10.15(a). Eddy currents are induced in the ring which acts as a short-circuited secondary of a transformer. The ring sets up a flux which—by Lenz's law—opposes the main flux. There is thus a delay in time before the flux in the shaded portion reaches its maximum value. The primitive 2-phase flux so produced causes a shift in field which is sufficient to cut the rotor conductors enabling the shaft to turn. The motor is employed for small fans; speed control may be fitted (Fig. 10.15(b)).

10.15. Motor Regulations

(1) Efficient means must be provided for starting and stopping, which will consist of an isolator and, except for the very small sizes, a starter. These devices must be easily operated and readily accessible to the person operating the motor.

The Electricity (Factory Act) Special Regulations 1908 *and* 1944, *require that in every place in which a machine is driven by a motor there shall be means at hand for either switching off the motor or stopping the machine, if necessary to prevent danger.*

For the purpose of inspection and other work, a single means of isolation for a group of motors is permitted.

(2) Every motor must be provided with a means to prevent restarting due to an appreciable drop in voltage or supply failure, where unexpected restarting might be dangerous. In practice, this requires that the motor starter must be provided with some form of no-volt trip. This regulation may be relaxed where the circuit contains automatic control devices, such as thermostats, which are necessary for operating purposes, or where a dangerous condition might result from the failure of a motor to restart after a temporary interruption of the supply.

Again, referring to the danger of unexpected restarting should there be more than one method of manually stopping a motor, means must be provided which will prevent restarting until every stopping device has been reset.

(3) Motors rated above 400 W must have a starter incorporating overload protection. The supply undertaking should be notified where there is likely to be excessive starting current; due to the surge of single-phase motors they are normally limited to a power of 1500 W.

(4) The minimum rating of cables to be equal to the full-load current of the motor. Cables subject to surges caused by frequent starting and stopping may require their cross-sectional area to be increased.

(5) Where the starter contains means for excess-current protection, the rating of the fuse or circuit-breaker protecting the final sub-circuit which supplies the motor, may be twice that of the cable between the fuse or circuit-breaker and the starter, e.g. assuming a cable rating of 21 A, then fuse or circuit-breaker rating may be 42 A (Fig. 10.16). It should be noted that cable sizes may require to be increased where there is the possibility of excessive voltage drop.

(6) Sizes of cables for rotor circuits of slip-ring or induction commutators must be suitable for load and starting conditions, bearing in mind that such rotor currents may be many times more than the supply current.

Fig. 10.16. Possible relation of final sub-circuit fuse or circuit-breaker to cable rating.

EXERCISES

1. Describe with connexion diagram, one form of split-phase single-phase motor. Explain the action of this type of motor, and explain why phase splitting is necessary.

A 22·38 kW, 240 V, single-phase motor runs at full load at 0·72 power-factor lagging. If the efficiency of the motor at full load is 85 per cent, calculate the current taken from the supply. [C]

2. Describe with sketches, and connexion diagram, a brake test on a three-phase motor.

Calculate the brake power, the power-factor and the efficiency of a three-phase motor from the test readings which follow:

Speed of motor	1420 rev/min
Three-phase wattmeter	13 500 W
Voltmeter	415 V
Ammeter	24·4 A
Brake pulley diameter	0·28 m
Effective pull at circumference of pulley	280 N [C]

3. Describe the characteristics of any *three* of the following motors, and their main uses for industrial purposes: (*i*) 3-phase induction motors, (*ii*) synchronous motors, (*iii*) a.c. commutator motors, (*iv*) d.c. shunt wound motors, (*v*) series (universal) motors. [C]

4. Discuss the procedure you would follow in inspecting and testing a cage-rotor induction motor and star-delta starter for the purposes of preparing a report upon their condition.

Your answer should describe the reasons for the tests you suggest, the instruments required, and the methods of making the tests.   [C]

5. List the tests and checks you would consider necessary for the compilation of a report upon the condition of a large 3-phase slip-ring induction motor together with its associated control gear.   [C]

6. Give details of the alterations in connexions necessary to reverse the rotation of the following types of motors, and illustrate your answers with diagrams. (i) 3-phase induction motor, (ii) universal motor, (iii) d.c. compound-wound motor, (iv) single-phase induction motor with split phase starting.   [C]

7. (a) The nameplate of a cage-rotor induction motor indicates that the motor is a 415-V three-phase type, although the connexion box has 6 un-marked terminals in it. Describe how you would identify the terminals.

(b) Draw a full circuit diagram of the motor connected to a star-delta starter with under-voltage and over-current protection.   [C]

8. The driven speed of a machine is to be 300 rev/min and the machine is fitted with a V rope pulley of 0·5 m pitch circle diameter. The available supply is three-phase and neutral at 50 Hz and the only motor available is a 6-pole induction type which at the rated load has a 5% slip.

(a) What size pulley must be fitted to the motor shaft?

(b) Determine from the following table the number of vee-belts required for a 12 kW motor.

belt speed (m/min)	150	330	450	660	800	900	1000
power/belt (kW)	1·5	2·0	2·5	3·0	3·5	4·0	5·0

9. (a) Explain briefly why a single-phase induction motor requires a starting winding whereas a 3-phase induction motor is self-starting.

(b) Draw a circuit diagram of a single-phase capacitor-start and run induction motor and describe the operation of starting the motor.

(c) A 2·5-kW single-phase motor connected to a 240-V supply takes a current of 16 A when running at full load at a power factor of 0·8 lagging. Calculate the efficiency at this load.

Chapter 11

TRANSFORMERS

11.1. Basic Considerations

As will be seen from the a.c. power equation, $P = VI \cos \phi$, raising the voltage reduces the current for a given power and power factor. The whole of the national grid depends upon this basic fact, so that enormous power can be carried by relatively thin transmission lines. The transformer can as easily step-down as step-up the voltage and may be designed in a number of different forms for various functions. It may even be used neither to raise nor lower the voltage —the bathroom shaver-socket to BS 3052 simply employs a $1:1$ transformer for isolation purposes.

In essence, the transformer usually consists of two windings insulated from each other and linked by a common magnetic field. The core consists of high-permeability steel laminations (which may be about 0·3 mm thick); these may be lightly insulated from each other by a phosphate coating. The principle of operation of the simple transformer may be understood by reference to Fig. 11.1, where both primary and secondary windings are placed on the central limb (in practice, for safety, the lower voltage winding would be the one nearer the core).

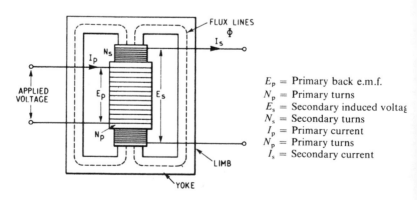

E_p = Primary back e.m.f.
N_p = Primary turns
E_s = Secondary induced voltag
N_s = Secondary turns
I_p = Primary current
N_p = Primary turns
I_s = Secondary current

Fig. 11.1. Simple double-wound transformer.

11.2. No-load Conditions

For simplification, the ideal case is considered, one in which all losses are neglected. When an alternating voltage is applied to the primary winding with the secondary winding open-circuited, the primary acts as a pure choke. A back e.m.f. (E_p) is produced which is opposite and equal to the applied voltage. A small current flows in the primary just sufficient to maintain the flux in the core, but the voltage drop in the primary winding due to it is so small as to be negligible. The alternating flux also embraces the secondary turns, therefore the change in flux linkages causes a secondary e.m.f. (E_s) to be induced across the secondary winding.

For perfect magnetic coupling and zero loss, the *ratio of primary and secondary e.m.f. is equal to the ratio of turns.*

The ratio of TRANSFORMATION,

$$\frac{E_p}{E_s} = \frac{N_p}{N_s}$$

By transposition,

$$\frac{E_p}{N_p} = \frac{E_s}{N_s}$$

so that the volts/turn of the primary and secondary are the same.

11.3. Load Conditions

When a load is connected to the secondary circuit a current I_s flows, a secondary magneto-motive force of value $I_s N_s$ ampere-turns is set up and current is drawn from the primary in the following manner:

The effect of I_s is to create a magnetic field which by Lenz's Law opposes the flux set up by the primary. Since the initial value of flux must be maintained because it gives rise to the back e.m.f., an additional current must circulate in the primary to produce an added value of magneto-motive force (ampere turns) to balance the secondary m.m.f. The primary back e.m.f. is equal and opposite to the applied voltage.

Thus it will be seen that the core flux remains constant through all load conditions.

Since ampere-turns balance

$$I_p N_p = I_s N_s$$

$$\frac{I_p}{I_s} = \frac{N_s}{N_p}$$

Ignoring losses, *the ratio of currents is inversely proportional to ratio of turns.*

11.4. E.M.F. Equation

The e.m.f. can also be expressed in terms of the maximum value of flux (Φ_m) and frequency (Hz).

If the flux wave is sinusoidal, the primary turns are cut by Φ_m Wb as the flux increases from zero to its maximum value and is repeated as the flux dies to zero. The turns are thus cut four times in each cycle, and this occurs f times per second

$$\text{Induced e.m.f.} = \text{flux cut/second}$$

$\therefore$ Average e.m.f. induced in the primary $= 4N_p f \Phi_m$

$$\text{Since form factor of a sine wave} = 1 \cdot 11 = \frac{\text{r.m.s. value}}{\text{average value}}$$

$$\text{R.M.S. value of primary e.m.f.} = 1 \cdot 11 \times \text{average value}$$

$$= 4 \cdot 44 N_p f \Phi_m$$

Similarly r.m.s. value of secondary e.m.f. $= 4 \cdot 44 N_s f \Phi_m$

Example 11.1

A 5:1 ratio transformer is worked at a maximum flux density of 0·5 T. If the cross-sectional area of the core is 20 cm² and the primary turns are 2000, calculate the secondary e.m.f. for a frequency of 50 Hz neglecting losses.

With a turns ratio of 5, the turns on the secondary winding are one-fifth of those on the secondary.

$$N_s = \frac{N_p}{5} = \frac{2000}{5} \qquad = 400 \text{ turns}$$

Now $$B_m = \frac{\Phi_m}{a}$$

$\therefore$ $$\Phi_m = B_m \times a = 0 \cdot 5 \times 20 \times 10^{-4} \qquad = 1 \text{ mWb}$$

Secondary e.m.f. $$E_s = 4 \cdot 44 N_s f \Phi_m$$

$$= 4 \cdot 44 \times 400 \times 50 \times 10^{-3} \qquad = \underline{88 \cdot 8 \text{ V}}$$

Example 11.2

A 500 kVA 6600/415 V transformer has 5000 turns on the primary. Calculate (a) the secondary turns (b) the approximate primary and secondary currents and (c) the maximum flux for a frequency of 50 Hz.

(a)

$$\frac{E_p}{N_p} = \frac{E_s}{N_s}$$

Secondary turns

$$N_s = N_p \times \frac{E_s}{E_p} = 5000 \times \frac{415}{6600} = \underline{315}$$

(b) $\qquad$ Current $= \dfrac{kVA}{V}$

∴ $\qquad$ Primary current $= \dfrac{500 \times 1000}{6600} \qquad = \underline{76\,A}$

$\qquad$ Secondary current $= \dfrac{500 \times 1000}{415} \qquad = \underline{1200\,A}$

$$E = 4{\cdot}44Nf\Phi_m$$

(c)

Maximum flux $\qquad \Phi_m = \dfrac{E}{4{\cdot}44Nf}$

Taking primary values

$$= \frac{6600}{4{\cdot}44 \times 5000 \times 50} \qquad = \underline{5{\cdot}95\,mWb}$$

11.5. Auto-transformers

Auto-transformers permit a reduction in cost as, instead of two separate windings, the secondary virtually forms a tapping off the primary (Fig. 11.2). Since the primary and secondary are in physical

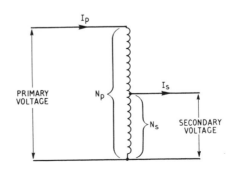

Fig. 11.2. The auto-transformer.

contact there are severe restrictions to its use (see the Regulations extract at the end of this chapter). A particular sphere of application where the restrictions do not apply is in motor starters when it is necessary to reduce the heavy starting current of certain motors.

The turns, voltage and current ratios are similar to the double-wound type.

Example 11.3

An auto-transformer has an output of 2 A at 8 V. If the secondary turns are 20 and the primary voltage is 240 V, calculate the number of turns in the primary winding and the current in the primary neglecting losses.

$$\frac{\text{Primary voltage}}{\text{Secondary voltage}} = \frac{\text{Primary turns}}{\text{Secondary turns}}$$

$\therefore$
$$\text{Primary turns} = 20 \times \frac{240}{8} \qquad = \underline{600}$$

$$\frac{I_p}{I_s} = \frac{N_s}{N_p}$$

$\therefore$
$$I_p = I_s \times \frac{N_s}{N_p} = 8 \times \frac{20}{600} \qquad = \underline{0\cdot267 \text{ A}}$$

11.6. Construction

Apart from mechanical protection the main purpose of design is to reduce the various losses, which although as a percentage are small may be of great magnitude in the larger transformers. Due to the fact that there are no moving parts, the dispersal of heat forms a special problem. There are two major classes of losses:

(1) *Copper or I^2R loss.* The former term is still used even though the use of aluminium conductors is on the increase. The wasted power appears as heat.

(2) *Iron or magnetic*

(*i*) EDDY-CURRENT LOSS. The alternating flux induces eddy currents in the mass of the core which also appears as heat. In order to reduce these small circulating currents, the core should possess a high ohmic resistance. The core is made up of laminations which consist of upright portions called the limbs, the yokes being the cross-pieces (Fig. 11.3). In one type the limb plates are securely

Fig. 11.3. Complete core and windings for 500 kVA, 3-phase, 1100/433 V
transformer, showing the h.v. connexions.

bound with cotton webbing. The yokes are held between heavy
angle or channel-section steel clamps which have tie rods at each end.
Up to 2500 kVA, it is more economical simply to overlap the
rectangular laminations at the corners. Above this size, mitred
corners are used and the limbs are bound at intervals by a number of
layers of epoxy resin, impregnated glass applied under tension and
subsequently cured.

(*ii*) LEAKAGE LOSS. Some of the flux from the primary windings
goes through air and does not link with the secondary, thus more
current is used than would otherwise be necessary. The transformer
illustrated in Fig. 11.3 is known as a shell-type transformer, which

has the advantage that the windings are almost completely sur-rounded by the magnetic steel. In the *core* type, the centre limb is removed and the windings are placed on the two outer limbs. Leakage is reduced by halving each winding so that half of the primary and secondary are positioned on each limb, the winding with the lower voltage being nearer the core from which it is insulated by a layer of pressboard (Fig. 11.4).

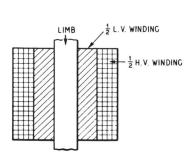

Fig. 11.4. Arrangement of concentric windings on each limb.

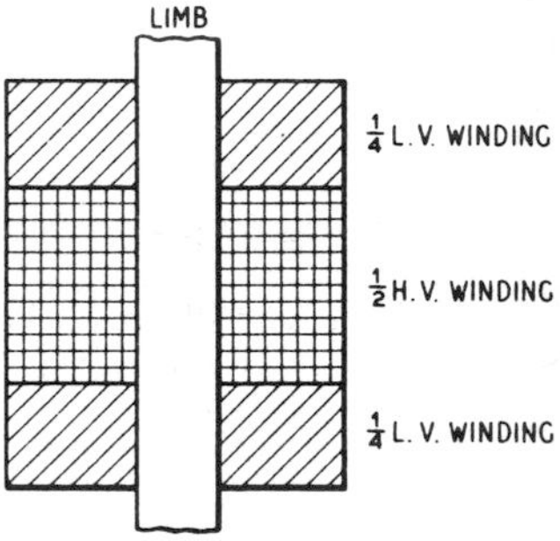

Fig. 11.5. Arrangement of sandwich winding.

(*iii*) HYSTERESIS LOSS. More current is required to demagnetise the core than to magnetise it. (See Section 1.21.) This represents a loss which is repeated every cycle and can only be reduced by careful selection of the core material.

Concentric windings are commonly used. These are cylindrical in form and are wound with wire, or strip copper in the larger sizes. At present the tendency is to replace copper by aluminium. The conductors are covered by either cotton or manilla paper or both. On the smaller sizes round enamelled wires may be used.

A sandwich type of winding may be adopted as an alternative to the concentric cylinder. Here the windings are arranged in layers (Fig. 11.5).

For increased efficiency 'grain-oriented' silicon iron is now used for the magnetic circuit. Here the hot-rolling process during manu-facture produces crystals of the material which point in one direction.

11.7. Cooling

In the smaller sizes up to a few kVA, natural air cooling is sufficient. Where excessive heat is experienced the windings are immersed in an oil-filled tank. Convection currents in the oil provides the cooling. The tanks are manufactured from mild steel plate with welded seams. Small power transformers up to about 25 kVA may have plain tanks and above this size, when the losses are too great to be dissipated by the sides of the tank, pipes or welded on radiators are fitted. A recent improvement is *Pyroclor*, which is a non-flammable synthetic liquid having an electric strength at least comparable with that of oil.

11.8. Phasor Diagram of Inductive Load

The phasor diagram for an inductive load on a transformer can be brought out by an example.

Example 11.4

A 400/200 V single phase transformer is supplying a load of 25 A at a power factor of 0·866 lagging. On no-load the current and power factor are 2 A and 0·208 respectively. Obtain graphically the current taken from the supply. The effects of resistance and reactance of the windings can be neglected.

Referring to Fig. 11.6, drawn to scale, the applied voltage V_p is drawn equal and opposite to E_1, the no-load current I_0 operating at a power factor of 0·208, i.e. a phase difference of 78°. The secondary current operates at a power factor of 0·866 and thus lags the induced secondary voltage E_s by 30°. With a 2:1 ratio the secondary current I_s is balanced by the primary current component I'_p. The phasor sum of I_0 and I'_p produces the primary current I_p of 14 A which lags V_p by 35°. Hence primary power factor is 0·82 lagging.

11.9. Transformer Regulation

Transformer regulation may be taken as the drop in secondary voltage as the secondary is loaded. The variation in secondary voltage between no-load and full-load is normally not more than $1\frac{1}{2}$ to 2%.

$$\text{Percentage regulation} = \frac{\text{No-load} - \text{full-load volts}}{\text{No-load volts}} \times 100$$

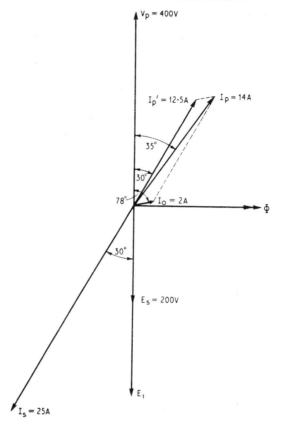

Fig. 11.6. Phasor diagram of inductive transformer load.

11.10. Prospective Short-circuit Current and Percentage Reactance

These terms occur when dealing with certain transformer calculations. Prospective short-circuit current is the current which can flow under short-circuit conditions. By modified Ohm's Law it may simply be obtained by dividing the combined impedance of the secondary winding and cables up to the point where the short-circuit occurs into the transformer secondary e.m.f.

Equivalent reactance of a transformer is often stated as a percentage instead of in ohms. If the voltage drop due to reactance is 6% of the secondary terminal voltage on no-load, then percentage reactance is 6. Where percentage resistance is relatively small only the percentage reactance need be taken for practical purposes.

Example 11.5

(a) *A transformer has a no-load and full-load secondary voltage of 500 V and 487 V respectively. Calculate the percentage regulation of the transformer.*

(b) *Two 3-phase generators and transformers are connected as in the following diagram. A short circuit across all phases occurs near the common busbars as shown. Calculate the short-circuit current.* [C]

(a) Regulation $= \dfrac{\text{no load sec voltage} - \text{full load sec voltage}}{\text{sec voltage}}$

$= \dfrac{500 - 487}{500} \times 100 \qquad \qquad = \underline{2 \cdot 6\%}$

(b) Converting percentage reactances to a common base of 2 MVA and working in MVA

No. 1. Generator $= \dfrac{6}{1 \cdot 5} \times 2$ $= 8\%$ reactance

No. 1. Transformer $= \dfrac{4}{2} \times 2$ $= 4\%$ reactance

TOTAL NO. 1. $= 8 + 4$ $= \underline{12\% \text{ reactance}}$

No. 2. Generator $= \dfrac{10}{1} \times 2$ $= 20\%$ reactance

No. 2. Transformer $= \dfrac{3 \cdot 5}{2} \times 2$ $= 3 \cdot 5\%$ reactance

TOTAL No. 2. $= 20 + 3 \cdot 5$ $= \underline{23 \cdot 5\% \text{ reactance}}$

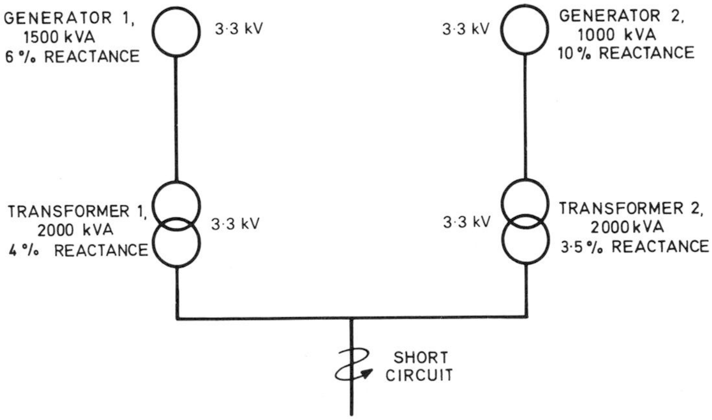

GENERATOR 1,
1500 kVA
6 % REACTANCE
3·3 kV

3·3 kV
GENERATOR 2,
1000 kVA
10 % REACTANCE

TRANSFORMER 1,
2000 kVA
4 % REACTANCE
3·3 kV

3·3 kV
TRANSFORMER 2,
2000 kVA
3·5 % REACTANCE

SHORT CIRCUIT

Fig. 11.7. See Example 11.5.

For parallel connection and with x as entire system percentage reactance

$$x = \frac{12 \times 23 \cdot 5}{12 + 23 \cdot 5}$$

$$\therefore \quad x = 7 \cdot 94\%$$

$$Short\text{-}circuit \ MVA \quad = 42 \times \frac{100}{7 \cdot 94} \qquad = 25 \cdot 18 \ MVA$$

$$Now \ MVA = \sqrt{3}VI$$

$25 \cdot 18 = \sqrt{3} \times 3 \cdot 3 \times 10^3 \ I$, where I = short circuit current

$$\therefore \quad I = \frac{25 \cdot 18 \times 10^6}{\sqrt{3} \times 3 \cdot 3 \times 10^3} \qquad = 4406 \ A$$

Example 11.6

The single-line diagram shows how a motor, having a full load input of 100 kVA is connected to a 11 kV supply. The rating and percentage

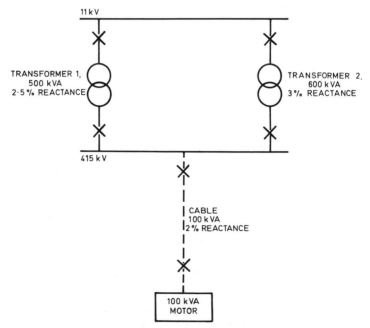

Fig. 11.8. See Example 11.6.

reactance of the transformers and the cable at full load is as shown.
Neglecting resistance and the internal impedance of the supply,
calculate:
   (a) *the required breaking capacity for circuit-breaker A;*
   (b) *the required breaking capacity for circuit-breaker B;*
   (c) *the voltages at A and B when the motor is taking a starting*
current of six times full-load current (assume that transformer No. 2 is
out of commission).

(a) Converting percentage reactances to a common base of
3000 kVA

$$\text{Transformer No. 1} = \frac{3000}{600} \times 2\cdot5 = 15\%$$

$$\text{Transformer No. 2} = \frac{3000}{500} \times 3 = 15\%$$

Resultant percentage reactance of two parallel transformers with
equal percentage reactances $= 7\cdot5\%$.
With prospective short-circuit current (defined as the current under
short-circuit conditions).

$$= \frac{100}{\text{percentage reactance}} \times \frac{\text{base kVA}}{10^3} \text{ MVA}$$

$$= \frac{100}{7\cdot5} \times \frac{3000}{10^3} \qquad\qquad = \underline{40\,\text{MVA}}$$

(b) At point B the reactance of the cable to be included to a base
of 3000 kVA, reactance of cable

$$= \frac{3000}{100} \times 2$$
$$= 60\%$$

Resultant reactance of the two transformers in parallel and the cable
in series

$$= 7\cdot5 + 60$$
$$= 67\cdot5\%$$

$$\text{Value of prospective short circuit} = \frac{100}{67\cdot5} \times \frac{3000}{10^3} \qquad = \underline{4\cdot5\,\text{MVA}}$$

(c) When starting load equals $6 \times 100 = 600$ kVA voltage drop in transformer No. 1

$$= \frac{600}{500} \times 2{\cdot}5 = 3\%$$

$\therefore$ voltage at A = 97% of 415   = <u>404 V</u>

Volt drop in cable

$$= \frac{600}{100} \times 2 = 12\%$$

Total voltage drop in cable

$$= 3 + 12 = 15\%$$
$\therefore$ voltage at B   = 85% of 415 V   = <u>353 V</u>

Example 11.7.

A 415 V 3-phase transformer is rated at 750 kVA and 4·75 percentage reactance.
Calculate (a) rated current and (b) prospective current.

(a) Rated current $= \dfrac{\text{kVA} \times 10^3}{\sqrt{3} \times V}$

$$= \frac{750 \times 10^3}{\sqrt{3} \times 415} \qquad\qquad = \underline{1043\ A}$$

(b) Prospective current

$$= \frac{\text{kVA} \times 10^3 \times 10^2}{\sqrt{3}V \times \text{percentage reactance}}$$

$$= \frac{750 \times 10^3 \times 10^2}{\sqrt{3} \times 415 \times 4{\cdot}73 \times 10^3} \qquad = \underline{22\ kA}$$

(If cable resistance of 0·0152/phase is added the fault current is reduced to about 14 kA).

11.11. Open-circuit Test

With the transformer run at the rated voltage and frequency and with the secondary open-circuited, the wattmeter in the primary circuit reads the no-load losses (Fig. 11.9); as the copper loss is negligible, the no-load losses are taken as the iron loss due to

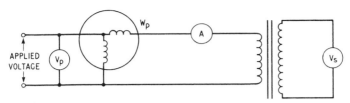

Fig. 11.9. Circuit for open-circuit test.

hysteresis and eddy currents and this is constant for all loads. The
no-load power factor may be obtained in conjunction with the
ammeter. V_p/V_s gives the turns ratio.

11.12. Short-circuit Test

The applied voltage is reduced so that the ammeter A_s (Fig. 11.10),
which is across the secondary winding, reads full-load current at
normal frequency. The primary voltage is so low and the flux so small
that the iron losses can be neglected. Under these conditions the
ammeter A can be taken as reading the copper loss.

11.13. Transformer Efficiency

The total losses are the sum of the iron and copper losses, as
obtained by the open-circuit and short-circuit tests.

$$\text{Efficiency} = \frac{\text{Output}}{\text{Input}} = \frac{\text{Output}}{\text{Output} + \text{losses}}$$

$$= \frac{\text{Output}}{\text{Output} + \text{iron loss} + \text{copper loss}}$$

Further, the maximum efficiency of a transformer occurs when the
copper losses are equal to the iron losses.

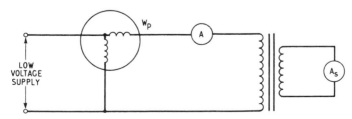

Fig. 11.10. Circuit for short-circuit test.

Example 11.8

(a) *Why is it usually necessary to cool transformers? Describe briefly two methods by which this can be done.*

(b) *A 20-kVA transformer when tested was found to have 600 W iron losses and 700 W copper losses when supplying full load at unity power factor.*

Calculate the efficiency of the transformer at unity power factor (i) on full load (ii) on half full load. [C]

(a) Explanation already given in text.

(b) (i) *Full load.* At unity power factor kW = kVA

$$\text{Efficiency per cent} = \frac{\text{Output}}{\text{Output} + \text{losses}} \times 100$$

Working in kW

$$= \frac{20}{20 + 0 \cdot 6 + 0 \cdot 7} \times 100 \qquad = \underline{93 \cdot 89\,\%}$$

(ii) *Half full load*

$$\text{Copper loss} = I^2 R \quad \text{i.e.} \propto \text{current}^2$$

$$= 700 \times (\tfrac{1}{2})^2 \qquad\qquad = 175\,\text{W}$$

$$\text{Efficiency} \% = \frac{10}{10 + 0 \cdot 6 + 0 \cdot 175} \times 100 \qquad = \underline{92 \cdot 8\,\%}$$

11.14. I.E.E. Regulations Governing Transformers

Control of a transformer must be by a linked switch for the purpose of isolating the transformer from all circuit conductors of the supply.

Where oil-filled transformers of a capacity more than 25 litres are installed, means must be provided for draining away surplus oil and preventing the escape of oil to any other part of the building.

In a Note to the Regulations, further precautions are given: (i) fitting of a drainage pit 'to collect leakages of liquid and ensure their extinction in the event of a fire' or (ii) 'installation of the equipment in a chamber of adequate fire resistance and the provision of sills or other means of preventing burning liquid spreading to other parts of the building, such chamber being ventilated solely to the external atmosphere.'

Auto transformers with an extra-low-voltage output shall not be used to supply:

(i) any socket-outlet;

(ii) any portable appliance, e.g. electric model railways and electric toys, unless containing the auto-transformer;

(iii) earth concentric wiring;

(iv) any e.l.v. circuit which is outside the scope of the Regulations, e.g. an electric bell circuit operating at a voltage not exceeding 15 V.

EXERCISES

1. Describe briefly two methods of cooling transformers. Explain why different methods are used for small and large transformers.

A 30 kVA single-phase transformer was known to have the following losses at full-load unity power-factor: iron losses, 500 W; copper losses, 600 W. Calculate the efficiency of the transformer at unity power-factor (*a*) on full load, (*b*) on half full load. [C]

2. (*a*) Explain the differences between a single-phase double-wound transformer and a single-phase auto-transformer. What are the requirements of the I.E.E. Regulations regarding the use of auto-transformers?

(*b*) A 40 kVA single-phase transformer was tested for efficiency by the 'open-circuit' and 'short-circuit' tests. On short-circuit, at full-load current, the power used was 1140 W. On open-circuit, the power used was 800 W. Calculate the efficiency of the transformer at unity power factor on (*i*) full load, and (*ii*) half full load. [C]

3. (*a*) What is the difference between a double-wound and an auto-transformer?

(*b*) What are the requirements of the I.E.E. Regulations with regard to the use of auto-transformers?

(*c*) A transformer of 500-kVA rating has iron losses of 2500 W and copper losses at full load of 4000 W. Sketch a graph showing how the total losses vary from no-load to full-load. What are the conditions for maximum efficiency? [C]

4. (*a*) Explain what is meant by the term 'prospective short-circuit rating' as applied to a consumer's installation.

(*b*) A consumer's low-voltage switchboard is fed by three 3000 kVA/415 V transformers connected in parallel, each transformer having a 4% reactance. The interconnection between each transformer and its associated low-voltage circuit-breaker consists of cables having a 2% reactance at 11 000 kVA. Ignoring the effect of the supplies up to the 11 kV transformers, find the maximum short-circuit MVA and the current which will flow in the event of a short-circuit fault on the busbars of the low-voltage switchgear.

(*c*) Find the maximum short-circuit current in (*b*) above if the low-voltage switchboard was replaced by three separate low-voltage switchboards, each fed by a 2000 kVA transformer.

5. Explain how the following types of single-phase transformer differ from the ordinary double-wound single-phase voltage transformer: (a) auto-transformer, (b) current transformer. For each case make a diagram showing the transformer in use and explain why it is so used. [C]

6. (a) What is an auto-transformer?

(b) Illustrate by diagram the windings of an auto-transformer, indicating the ratio of output to input pressure as shown on your drawing.

(c) What restrictions do I.E.E. Regulations impose on the use of these transformers? [C]

Chapter 12

POWER-FACTOR IMPROVEMENT—TARIFFS

12.1. Advantages of High Power Factor

Elementary considerations of the simple power equation bring out the disadvantages of a low power factor. The expression $P = VI \cos \phi$ was referred to when we dealt with transformer theory; here we are concerned with $\cos \phi$.

For a fixed voltage, a reduced power factor requires a *larger current for a given power*. This in turn leads to increased IR voltage drops and I^2R losses in transmission. For these reasons, supply undertakings offer certain reduced tariffs as an incentive for power-factor correction since, in addition, the generator capacity is limited by low power factors of consumers' plant.

Example 12.1

What is the maximum output of a 6600-V 1000-kVA alternator feeding loads at 0·65 power factor?

$$\text{Current} = \frac{\text{kVA}}{\text{V}} = \frac{1000 \times 1000}{6600} = 151\cdot5\,\text{A}$$

Maximum power output $= VI \cos \phi$

$$= \frac{6600 \times 151\cdot5 \times 0\cdot65}{1000} = \underline{650\,\text{kW}}$$

A greatly increased generator plant capacity would be required for the 1000 kW output. At unity power factor kVA = kW, so that the same generator would deliver this output when $\cos \phi = 1$. Poor power factors incur even further costs as switchgear and cables are rated in amperes.

12.2. Causes of Low Power Factor

Low power factors are produced by apparatus such as induction motors, especially at low loads, and the ballast units of discharge lighting which need a magnetising reactive current for their action; electric arc welders also have a low power factor. The magnetic field of such apparatus necessitates a current which does no useful work and does not result in heat or mechanical power but is simply required to build up the field. Although the current is returned to the mains as the field collapses, added cable section and plant becomes necessary to carry this current. Only the active or useful component of the current is responsible for the useful work done by the apparatus.

Another way of looking at this problem is to realise that a poor power factor causes the voltage and current to be out of phase so that their product does not produce power in watts but volt-amperes. An excess of capacitor plant would also bring about a poor power factor but it is unusual for this to occur.

The basic principle of power factor improvement is to inject a leading current into the circuit so as to neutralise the effects of the lagging magnetising current. Under these conditions the energy required by the magnetic field circulates between the correction apparatus and the corrected installation instead of between the installation and the supply. Since the cost of correction plant increases as unity p.f. is approached it is normal to limit the improvement to 0·95.

12.3. P.F. Improvement Plant

Where the problem of power-factor correction exists, improvement is effected by the installation of static capacitors or synchronous motors. Phase advancers have sometimes been used for this purpose. These methods may be costly so that, wherever possible, equipment should be purchased that operates at an inherently high power factor. Fluorescent fittings invariably contain capacitors for power factor improvement. Where a factory possesses a number of induction motors, the choice has to be made as to whether individual or centralised improvement is preferable. The obvious decision is to place a capacitor across the terminals of each induction motor but this may be uneconomical where many such motors are involved. Because the power factor of an individual motor may vary with the load—being lowest at low load—the capacitor may result in over-correction at certain loads and even produce a voltage surge that can have a damaging effect on the motor.

Example 12.2

The total current taken by a number of identical single-phase motors is 200 A with a power factor of 0·5, when their capacitors are out of circuit. Calculate:

(a) What the total current will be when the power factor is brought to unity by the switching in of the capacitors.

(b) What both the total current and power factor will be when 50% of the motors are switched off with the capacitors still in circuit.

Illustrate the above calculation by means of a phasor diagram of each of the three conditions.   [C & G]

(a) $\cos \phi = 0.5$

∴ $\phi = 60°$

Referring to triangle OAB (Fig. 12.1(a)), i.e. when the capacitors are out of circuit,

Reactive current $AB = I_m \sin \phi = 200 \times 0.866 = 173.2$ A

This has to be neutralised by an equal and opposite leading current OC to produce unity power factor represented by OA.

Current taken at unity power factor,

$$I = I_m \cos \phi = 200 \times \tfrac{1}{2} \qquad = \underline{100 \text{ A}}$$

(b) Referring to Fig. 12.1(b):

With 50% of motors $I_m = 100$ A

Phase angle as before

$$\phi = 60°$$

Active component of motor load

$$OA = I_m \cos 60° \qquad\qquad = 50 \text{ A}$$

Reactive component of motor load

$$AB = I_m \sin 60° \qquad\qquad = 86.6 \text{ A lagging}$$

Reactive component with capacitors in circuit

$$= 173.2 - 86.6 \qquad\qquad = 86.6 \text{ A leading}$$

New current, $I = \sqrt{(86.6^2 + 50^2)} \qquad = \underline{100 \text{ A}}$

New phase angle $\cos \phi = \dfrac{OA}{OC} = \dfrac{50}{100} \qquad = \underline{0.5 \text{ leading}}$

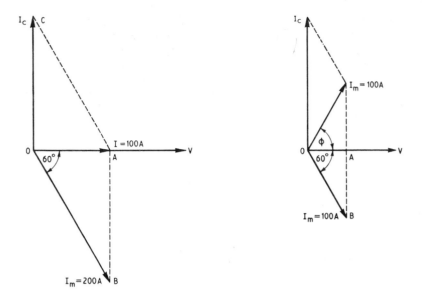

Fig. 12.1. (*a*) Motor current corrected to unity power factor. (*b*) Motor current over-corrected.

The effects of uneven correction may be overcome by automatic power-factor correction plant. This may employ a special relay to operate a reversing motor which drives a camshaft actuating mercury switches to control a number of capacitors through the medium of contactors.

Where there are very many units requiring improvement, central correction is normally preferable. Alternatively, individual correction may be applied to large motors and to group the smaller motors together for collective improvement.

12.4. Static Capacitors

Static capacitors are commonly employed for power-factor correction. Due to the absence of moving parts, installation and maintenance are comparatively simple. Other advantages are that capacitors are reliable and little or no floor space is necessary. The apparatus may be mounted on stanchions or walls where floor space is not available.

The capacitors consist of elements connected in parallel. For 3-phase work, capacitors are connected in delta. If connected in star, the voltage across each capacitor is reduced to $V/\sqrt{3}$ (*V* being supply voltage), and since the capacitance required is inversely proportional to V^2, the total capacitance required is three times as much as would be required with capacitors connected in delta. Therefore delta connexion is preferable if the capacitors are designed for the higher working voltage. Each individual element of a capacitor consists of two metal foils, either aluminium or tin, interleaved with high-quality tissue papers and wound into a common roll. The voltage at which the element is to operate governs the number and thickness of the tissue papers between the foils. During winding, a number of foil connexion tabs making contact with each foil, are inserted at regular intervals, to which after bunching, the necessary connecting wires are either soldered or welded. For the smaller ratings up to 250 V, the capacitor elements are impregnated with petroleum jelly. Oil impregnation is used for the majority of larger power-factor-correction capacitors. In respect of oil-filled tanks the Regulations as stated for transformers must be observed. The capacitors must also be supplied with an appropriate discharge resistor in order to ensure automatic discharge as the supply is disconnected (see

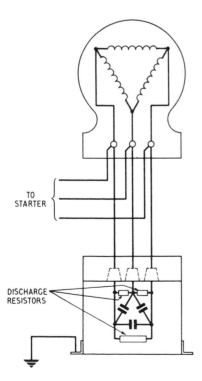

TO
STARTER

DISCHARGE
RESISTORS

Fig. 12.2. 3-phase power-factor improvement capacitors connected to induction motor stator.

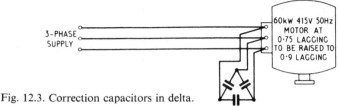

Fig. 12.3. Correction capacitors in delta.

Fig. 12.2, which also shows the connexions to the stator of a 3-phase induction motor).

Example 12.3

The power taken by a 415 V 50 Hz three-phase motor is 60 kW at 0·75 power factor lagging. A bank of capacitors is connected across the supply to improve the power factor.

Calculate the capacitance per phase and the total capacitance, required to raise the power factor to 0·9 lagging. [C]

The facts of the question may be summarised as in Fig. 12.3.
Referring to phasor diagram (Fig. 12.4),

Phase angle corresponding to 0·75 lagging = 41° 25′

Phase angle corresponding to 0·9 lagging = 25° 51′

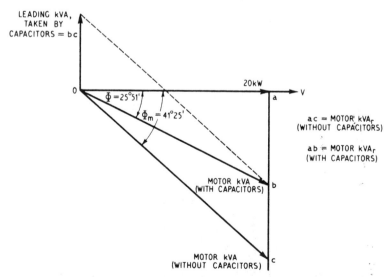

Fig. 12.4. Phasor diagram for circuit problem taking phase values.

Working in phase values:

$$\text{Motor power} = \tfrac{60}{3} \qquad\qquad = 20 \text{ kW}$$

To bring to 0·9 p.f., reactive component bc must be neutralised by equal and opposite leading kVAr

Taking triangle Oac,

$$ac = 20 \tan 41° 21'$$

$$ab = 20 \tan 25° 51'$$

$$bc = ac - ab$$

$$= 20 \tan 41° 21' - 20 \tan 25° 51'$$

$$= 20 (0·8801 - 0·4845)$$

$$= 20 \times 0·3956 \qquad\qquad = 7·912 \text{ kVAr}$$

Capacitor current

$$I_c = \frac{\text{kVAr}}{V} = \frac{7·912 \times 1000}{415} \qquad = 19·06 \text{ A}$$

Capacitance reactance

$$X_c = \frac{V}{I_c} = \frac{415}{19·06} \qquad\qquad = 21·76 \ \Omega$$

Hence

Capacitance required/phase

$$= \frac{10^6}{X_c 2\pi f} \quad (\text{since } X_c = 10^6/2\pi f C)$$

$$= \frac{10^6}{21·76 \times 314} \qquad\qquad = 146 \ \mu\text{F}$$

When connected in delta,

Total capacitance required

$$= \sqrt{3} \times \text{phase capacitance}$$

$$= \sqrt{3} \times 146 \qquad\qquad = 253·3 \ \mu\text{F}$$

12.5. Synchronous Motor

This machine can be run at a high power factor and thus may be employed instead of a 3-phase induction motor; unfortunately it is much more expensive. By increasing the d.c. exciting current, the power factor can be made to lead and so offset the lag of other machines. Synchronous motors may thus be employed for power-factor correction.

In the d.c. shunt motor an increase in load torque brings a speed decrease so that the back e.m.f. decreases. The resultant voltage becomes larger bringing about a corresponding increase in the load current. In the synchronous motor the speed is constant so that the back e.m.f. is constant; the latter can thus only be changed by varying the d.c. excitation.

By reference to the simplified phasor diagram (Fig. 12.5), the back e.m.f. E is almost equal and opposite to the applied voltage V. The phasor resultant V_R is almost in quadrature to the load current I. Changing the exciting current varies the magnitude of the back e.m.f. so that the phase relationship between V and I can be varied by this means. Sometimes synchronous motors are not used to drive

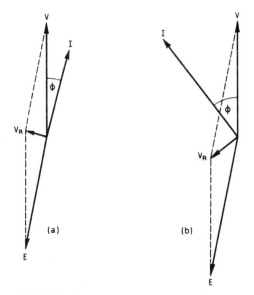

Fig. 12.5. Power factor improvement by synchronous motor. (a) Lagging power factor. (b) Leading power factor by an exciting current increasing back e.m.f.

V = supply voltage I = motor current
E = back e.m.f. ϕ = phase angle
V_R = resultant voltage

any machinery but simply as a method of correcting the p.f. of other apparatus; they then become known as *synchronous capacitors.*

12.6. Phase Advancer

The phase advancer is yet another method of manipulating the phase angle in order to effect power-factor improvement. Here it is applied to large induction motors. The phase advancer is a form of a.c. exciter which is mounted on the motor shaft, the a.c. exciter being connected so as to inject current into the induction-motor rotor circuit at slip frequency. By providing magnetising ampere-turns in the rotor in place of the supply magnetising ampere-turns, it neutralises the lagging component of the stator current thereby raising the power factor.

12.7. Tariffs

Electricity tariffs are the charges that are passed on to the consumer in order to pay for the generation and distribution of electrical energy. Capital costs must also be included to pay for the installation of new plant. Due to the variety of ways in which electricity is used, the charges may take many different forms in order to encourage the best utilisation of the generated electricity.

FLAT RATE. The chief merit is its simplicity although in order to preserve an adequate load factor, lighting and power are metered separately and the latter charged at a lower rate. This system is rapidly disappearing. The term 'load factor' is usually applied to a 24-hour period and may be expressed as a percentage, being equal to $\dfrac{\text{average kW}}{\text{maximum kW}} \times 100$ for this period.

TWO-PART DOMESTIC. Under the two-part or 'all-in' tariff the charge to the customer comprises two elements which are related, in theory at least, to the fixed costs and running costs of the supplies.

Area Boards themselves purchase electrical energy from the Central Electricity Generating Board, the supply being metered at the bulk supply points. The Boards recover their costs by charging consumers at an economic amount arrived at by reasonably simple tariffs. Tariffs are based on the fact that supply costs are generally of two kinds, (a) standing and (b) running.

Standing or fixed costs go on all the time regardless of the amount of power supplied. They consist of salaries and wages, rents, interest

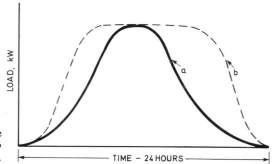

Fig. 12.6. With the same peak loads, consumer *b* has a better load factor.

on capital expenditure, insurance etc. By law, a supply must be available at all times and it follows that the generation and distribution costs per electrical unit (kWh) must be higher for a load used for a short period than for a similar load supplied for a longer period—because the standing costs remain the same.

The curves in Fig. 12.6 illustrate the power loads taken by two consumers *a* and *b*. Consumer *a* has a low load factor, while consumer *b* shows a reasonably well-sustained load which makes for efficient use of the electrical generating and distribution equipment. The area under the curves represents the number of units used since it is equal to kW × time in hours.

The domestic fixed charge sometimes appears to be arrived at somewhat arbitrarily. According to locality, it may be based on the number of rooms, floor area, size of premises rateable value or an assessed sum per week.

Running charges are very largely based on the costs relating to the fuel consumed by the generators. Nevertheless there is some variation between the kind of tariff and the unit charge as offered by different supply undertakings.

BLOCK TARIFF. The block tariff is largely supplanting the straight domestic two-part tariff as outlined above, being considered a fairer means of charging whilst still encouraging maximum energy utilisation. This tariff obviates the need for a fixed charge.

An initial fixed number of units ('a block') are priced at a relatively high price per unit, all other units consumed being at a cheap rate; there may also be succeeding blocks. The following figures might be quoted as a typical tariff:

> First block of 546 units at 15p per unit
> Second block of 5460 units at 6.9 p per unit
> Remainder at 3·75p per unit

Commercial and industrial block tariffs operating the same principle tend to be more involved.

NIGHT AND DAY RATE. This is one of the newer rates offered by some undertakings. A cheaper than normal unit charge is made for each unit supplied between 9 p.m. and 7 p.m. (10 p.m. and 8 a.m. during summer time). At other times the running charge is increased and there is also a 30% increase in the fixed charge. The tariff is advocated where more than one-fifth of the units are used at night time.

SLOT-METER METHODS. Most rates employ credit meters for measuring the number of units consumed and meter rents are nowadays rarely demanded. With prepayment meters however there may be a meter charge. Ingenuity has been displayed in avoiding having to make this somewhat unpopular charge, one simple alternative being to include a 20% increase in the fixed charge.

Prepayment methods may include the variable block tariff but the meters themselves tend to be more complicated. They are then often termed stepped-rate meters; when the first block of units have been consumed, the meter automatically changes over to a lower rate.

12.8. Off-peak Electricity

Cheaper rates are offered at night time, say between 10 p.m. and 7 a.m., and may apply to all types of premises. Electricity consumed during these hours obviously improves the load factors and brings down the price of generating and distributing each kilowatt-hour. The saving is passed on to the consumer.

Cheap rates have enormously encouraged the installation of *storage* water-heating and space-heating schemes, originally in commercial and industrial buildings and more recently in domestic premises. It is also sometimes applied to battery-charging circuits and pumped-water storage operations. Control of the off-peak periods is normally by time-switches. In the larger schemes 'ripple control' is sometimes used, a series of a.c. impulses at selected frequency (400–1500 Hz) being sent over the distribution lines with the object of opening and closing relays set to resonate at these frequencies.

Storage circuits (unless operating on time-of-day tariff) must be wired so as to be completely separate from the normal unrestricted supply. A further point to be noted is that no diversity should be allowed in calculating the total load.

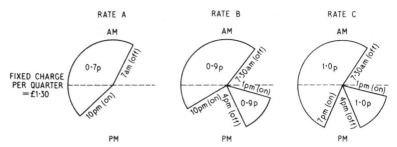

Fig. 12.7. Variations in off-peak rates as offered by one supply undertaking.

Variations in off-peak tariffs are typified by a supply undertaking's A, B and C rates as summarised in Fig. 12.7. Rate B includes a 'mid-day boost' and rate C gives a longer off-peak period at an increased running charge.

12.9. Maximum Demand

As electricity cannot be stored to any large extent, peak loads for short periods are responsible for increasing generating costs.

The rates for commercial and industrial consumers who use big supplies of electricity are directly related to (a) kWh and (b) maximum demand. M.D. meters are installed in order to record the kVA of maximum demand over half-hour periods (in some areas the actual time of maximum demand may determine the charge).

As we have seen, low power factors also increase the generating and distribution costs especially when they occur at a period of maximum demand. Since $kVA = kW/p.f.$, an increase in the power factor from 0·5 to unity cuts the kVA to a half of its former value. Authorities may exact penalties for power factors below 0·9–0·95 lagging.

The running charge will also be affected by a fuel adjustment clause for each penny by which coal fuel per ton costs more or less than a specified figure.

12.10. Worked Examples

Example 12.4

An installation comprises two 3-kW electric fires, eight 75-W lamps and ten 100-W lamps. It is estimated over a year that the average use of both fires will be two hours daily and of the lamps 25 hours weekly. Calculate

   (i) the annual cost with each of the tariffs given below and
   (ii) the average price per unit with tariff (b).

Tariff (a), flat rate, 3p per unit for heating and 7·5p per unit for lighting; tariff (b), all-in rate, £13·50. per annum plus 1·5p. per unit for all purposes.

(i) Annual heating load
$$= kW \times \text{hours used}$$
$$= (2 \times 3) \times (2 \times 365) \qquad\qquad = \underline{4380\ kWh}$$

(ii) Annual lighting load
$$= (8 \times 75 + 10 \times 100) \times (25 \times 52) \quad = 2080\ kWh$$

UNDER FLAT-RATE TARIFF (a)

Annual cost of heating at
3p per unit = 4380 × 3p $\qquad\qquad = £131\cdot40$

Annual cost of lighting at
7·5p per unit = 2080 × 7·7 $\qquad\qquad = \underline{£156\cdot00}$

$$\text{Total annual cost of tariff } (a) = \underline{£287\cdot40}$$

UNDER ALL-IN TARIFF (b)

Total units $\quad = 4380 + 2080 \qquad\qquad = 6460\ kWh$

Cost of 6460 units at 1·5p per unit $\qquad = £96.90$

Fixed charge $\qquad\qquad\qquad\qquad\qquad = \quad \underline{13\cdot50}$

$$\text{Total annual cost of tariff } (b) \quad £110\cdot40$$

(ii) Average price per unit under tariff (b)

$$= \frac{\text{total cost}}{\text{total units}}$$

$$= \frac{114\cdot40}{6460} \qquad\qquad = \underline{1\cdot7p}$$

Example 12.5

A consumer has a lighting load of 3·6 kW and is to install electric heaters. Alternative tariffs are available; two-part, £2·50 per quarter plus 0·7p for all units consumed, and a flat rate 2p per unit for lighting and 0·5p for heating.

Assuming all apparatus will have an average use of four hours daily throughout the year, calculate the kW rating of the proposed heaters so that the annual cost will be the same under either tariff.

Two-part Tariff

Running charge = total load (kW) × annual hours of use

$$\times \text{cost/unit}$$

Let x = kW rating of heaters

Running charge = $(3 \cdot 6 + x) \times (365 \times 4) \times 0 \cdot 7p$

$$= (3679 + 1022x)p$$

Fixed annual charge = £2·50 × 4   = 1000p

Total charge = $(3679 + 1022x + 1000)p$

$$= (4679 + 1022x)p \quad (1)$$

Flat-rate Tariff

Lighting cost = load × hours × cost/unit

$$= 3 \cdot 6 \times 4 \times 365 \times 2 = 10512p$$

Heating cost = $x \times 4 \times 365 \times 0 \cdot 5 = 730xp$

Total charge = $(10512 + 730x)p \quad (2)$

Since the annual costs under the two tariffs are to be the same, expression (1) must be equal to expression (2):

$$\therefore \quad 4679 + 1022x = 10512 + 730x$$

$$1022x - 730x = 10512 - 4679$$

$$292x = 5833$$

$$x = \frac{5833}{292} = 19 \cdot 97 \text{ kW}$$

Therefore for equal total annual cost under the two tariffs, the heating load must be 19·97 kW, or 20 kW in round figures.

Example 12.6

The tariffs available for supply to a factory are:
(a) Installed Load Rate
     Per quarter per kVA of: lighting 50p; cooking and water heating 25p; space heating 60p.
     Per quarter per kW of motors, 50p.
     Plus unit charge of 0·77p.
(b) Maximum Demand Rate
     £12 per annum for each of the first 20 kVA of recorded maximum demand.
     £10·50 per annum for each of the next 180 kVA of recorded maximum demand.
     Plus unit charge of 0·44p.

Installed loads and estimated usages were as follows:

Lighting	40 kVA and 45 000 *units per annum*
Cooking and water heating	30 kVA and 25 000 *units per annum*
Power	50 kVA and 80 000 *units per annum*
Space heating	20 kVA and 60 000 *units per annum*

Set out the cost lines for these tariffs and, assuming a maximum demand of 57 kVA, *determine:*

(*i*) *the most favourable tariff at the usage given,*

(*ii*) *the annual cost under each tariff with the usage of each section reduced by* 50%,

(*iii*) *at what annual usage the annual cost is equal for both tariffs.*

(i) (*a*) INSTALLED LOAD RATE

Fixed Charges

Lighting	40 kVA at £2.00 p.a./kVA	=	£80.00
Cooking and water heating	30 kVA at £1.00 p.a./kVA	=	£30.00
Power	50 kVA at £2.00 p.a./kVA	=	£100.00
Space heating	20 kVA at £2.40 p.a./kVA	=	£48.00
			£258.00

Running Charges

Total units = 45 000 + 25 000 + 80 000 + 60 000 = 210 000

210 000 units at 0·77p	=	£1617
Total fixed and running charges		£1875

(*b*) MAXIMUM DEMAND RATE

First 20 kVA of M.D. at £12.00	=	£240.00
Successive 37 kVA of M.D. at £10.50	=	£388.50
210 000 units at 0.44p	=	£924.00
		£1552.50

At the above usages tariff (*b*) *is cheaper*

(*ii*) (*a*) Running charge with 50% reduction in usage

£1617 ÷ 2	=	£808.50
Fixed charge unaltered	=	£258.00
		£1066.50

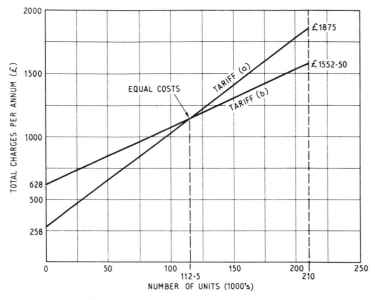

Fig. 12.8. Graphical solution to tariff problem.

(b) Running charge with 50% reduction

£924 ÷ 2	= £462.00
Fixed charge	= £628.50
	£1090.50

At these usages tariff (a) is cheaper

(iii) From the graph (Fig. 12.8) the crossover point, i.e. when the annual cost is equal for both tariffs, occurs at a usage of approximately 113 000 units. Calculation gives a more exact figure:

Let x = Number of units for equal costs then (working in pence)

$$258\,000 + 0.77x = 628.50 + 0.44x$$

$$x = \frac{37\,050}{0.33} = 112\,300 \text{ units}$$

Although the charges are dated due to inflation, the principles behind the calculations remain unaltered.

Example 12.7

A power consumer has an annual consumption of 860 000 units and maximum load of 450 kW at 0·75 p.f. The tariff applicable to this load is £6.00 per annum of kVA of maximum demand, plus 0·48p per unit.

Calculate the overall charge per unit. [C]

$$\cos \phi = \frac{kW}{kVA}$$

$$\therefore \quad kVA = \frac{kW}{\cos \phi} = \frac{450}{0·75} \qquad = 600$$

Charge for maximum demand

$$£6 \times 600 = £3600 \qquad\qquad = 360\ 000p$$

Running charge = $\qquad 860\ 000 \times 0·48 \qquad = 412\ 800p$

$$\qquad\qquad\qquad\qquad\qquad\qquad\qquad\qquad 772\ 800p$$

$$\text{Overall cost/unit} = \frac{772\ 800}{860\ 000} = 0·9p$$

12.11. White Meters

These have been primarily developed by Electricity Boards to boost the night use of electricity. The white-meter tariff has many economic advantages over plain off-peak tariffs. The white meter has two sets of cyclometer indicator dials, enabling any of the consumer's appliances to be used at any time during the day *or* night without any restrictions applying to cheap night rates.

The real saving comes where water is heated electrically during the night time together with long-term space heating, such as storage radiators, Electricaire, or floor warming, which can accept an 8-hour charge. Advantage could also be taken to carry out some of the electric cooking at night time. Thus in certain cases timed oven control has made off-peak cooking possible. By the use of the white meter, night electricity may be extended to such houshold chores as home laundry and dish-washing.

EXERCISES

1. State how you would explain power-factor and power-factor improvement to an electricity consumer.

The following instrument readings were taken of a single-phase load: voltmeter 240 V, ammeter 35 A, wattmeter 5710 W. Supply frequency is 50 Hz. Calculate (a) power factor, and (b) the value of the capacitance required in parallel to raise the overall power-factor to unity.   [C]

2. Describe the chief factors which govern the selection of (a) synchronous motors and (b) capacitors for the power-factor improvement of: (i) slip-ring motors, (ii) squirrel-cage motors, (iii) installations of small motors and appliances.

3. Explain, with a diagram, the meaning of power-factor correction. Describe one form of power-factor correction apparatus which may be installed in a large works. Clearly explain where it may be installed, its connexion and operation.   [C]

4. (a) Define power factor and illustrate your answer by a phasor diagram.

(b) The following values in each phase were recorded by a direct-reading instrument in a 3-phase, 3-wire circuit: voltmeter 415 V, ammeter 37·5 A, wattmeter 7570 W. Calculate the power factor of the circuit.

5. (a) The tariff for a 3-phase 415-V works installation is £10·00/kVA of maximum demand plus 0·4p/unit consumed.

(i) Calculate the annual cost for a maximum demand of 900 kVA and consumption of $1·5 \times 10^6$ units p.a.

(ii) Find the annual cost/unit.

(iii) Find the saving in annual cost if the power factor was raised from 0·8 lagging to 0·9 lagging with the same unit consumption.

(iv) Briefly explain what meters would be required for the above.

(b) Briefly justify the above type of tariff for large users of electrical energy.

6. The efficiency of a 415-V, 50 Hz, 746 kW, 3-phase motor, working at full load with a power-factor of 0·72 lagging is 86 per cent. What value of capacitance is necessary to raise the power-factor to unity? By the aid of a neat diagram, show how the capacitors would be connected.   [C]

Chapter 13

EARTHING AND EARTH-LEAKAGE PROTECTION

13.1. Protective Methods

Danger is defined in the Electricity Factory Acts (1908 and 1944) as meaning 'danger to health or danger to life or limb from shock, burn or other injury to persons employed, or from fire attendant upon the generation, transformation, distribution, or use of electrical energy'. It may be taken that all electrical safety regulations are designed to avoid such hazards.

Earthing of exposed metal parts is the main means adopted in this country against the effects of earth-leakage currents, although the I.E.E. recognises three other types of protection for apparatus and conductors:

(1) ALL-INSULATION. Enclosure by durable and substantially continuous insulation. A note maintains that if fixed apparatus, although not shrouded in insulating material, is nevertheless so guarded that it cannot be touched and cannot come into contact with exposed metal, it can in certain circumstances be regarded as coming within this category of all-insulated construction.

(2) DOUBLE-INSULATION. Appliances or lighting fittings having double insulation conforming to the appropriate British Standard. This simply refers to primary and secondary layers of insulation. Where double insulation is permitted as a recognised means of protection, appliances must comply with BS 3456 and bear the certificate mark of the British Electrical Approvals Board (B.E.A.B.), which are two concentric squares (Fig. 13.1(a)).

Other double-insulated apparatus must bear the certification mark of the British Standards Institution (Fig. 13.1(b)).

(3) ISOLATION. Isolation of metal so that it cannot come in contact with (a) live conductors or (b) earthed metal.

13.2. Protection against Short-circuit Currents

In dealing with transformers certain calculations have already been made. Here we are concerned with the maximum duration,

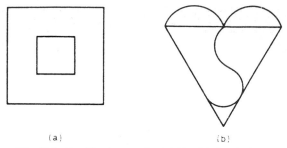

(a)   (b)

Fig. 13.1. Certification marks. (a) Double-insulation.
(b) British Standards Institution.

principally to limit fire risks; Fig. 1.15 indicates over-current
mechanical damage. Short-circuit over-current protective devices
must also have adequate breaking capacity in order to deal with
these abnormal currents.

Where the short-circuit time is limited to 5 s and there is a
conductor of cross-sectional area with a minimum of 10 mm², then
the following equation from I.E.E. Regulation 434–6, can be
applied:

$$t = \frac{k^2 a^2}{I^2} \text{ s}$$

where t = duration in s
a = cross-sectional area in mm²
I = a.c. short-circuit current

K = 115 for copper conductors insulated with p.v.c.
134 for copper conductors insulated with 60°C rubber, 85°C
rubber.
134 for copper conductors with 90°C thermosetting insulation.
108 for copper conductors insulated with impregnated paper.
135 for mineral-insulated cables with copper conductors.
76 for aluminium conductors insulated with p.v.c.
89 for aluminium conductors insulated with 60°C rubber,
85°C rubber.
94 for aluminium conductors with 90°C thermosetting
insulation.
71 for aluminium conductors insulated with impregnated
paper.
87 for mineral insulated cables with aluminium conductors.
100 for tin-soldered joints in copper conductors,
corresponding to a temperature of 160°C.

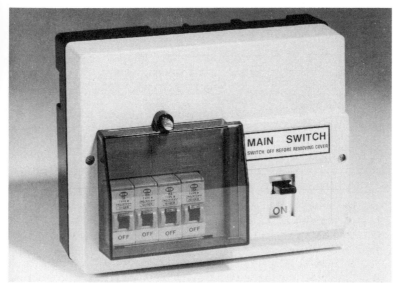

Fig. 13.2. 4-way MEMERA 21 moulded surface-mounting consumer unit complete with MCBs with a cut-off period within 0·01 s when fault current 2·7 to 4 times rating.

Example 13.1

(a) State the requirements for over-current protective devices.
(b) What is the maximum breaking time with a prospective fault current of 5000 A for
   (i)   25 mm² mineral insulated cable,
   (ii) 25 mm² p.v.c. insulated cable?

(a) To prevent danger, over-current protective devices must be able to act rapidly at values related to safe circuit current values. In addition they must be able to both make and break prospective short-circuit currents.

The 15th Edition repeats basic principles which go back to the early days of electric wiring, that these devices 'are suitably located and are constructed so as to prevent danger from overheating, arcing or the scattering of hot particles when they come into operation and to permit ready restoration of the supply without danger.'

The effects of electric shock are not only dependent upon the current flowing but also on the *time* this fault current will flow. To comply with the Regulations disconnecting time is given as 5 s for fixed equipment and only 0·4 s for apparatus fed from socket-outlets.

(b)(i)   Using $t = \dfrac{k^2 a^2}{I^2}$ s

and substituting   values $= \dfrac{135^2 \times 25^2}{5000^2}$ $= \underline{0·46 \text{ s}}$

(ii) $t = \dfrac{115^2 \times 25^2}{5000^2}$ $= \underline{0·32 \text{ s}}$

13.3. Principles of Earthing

The earthing of the consumer's installation must be carried out in such a manner that no fault of negligible impedance to earthed metal shall be so substained as to cause danger within the meaning of the word 'danger' as already defined.

The diagram in Fig. 13.3 shows a typical supply and installation much simplified. At the supply point, the neutral is connected to earth so that if a live conductor is inadvertently in contact with the earthed metal case, an alternative path is provided for the fault current.

In order to blow the fuse *the impedance of the earth circuit must be sufficiently low to pass the necessary current.*

Example 13.2

On a 2 kW kettle connected to a 240 V supply, a fault to earth occurs, through the earthed metal casing of the appliance, at a point a

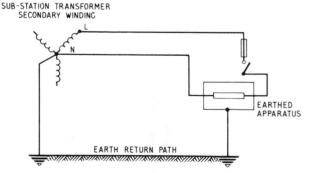

SUB-STATION TRANSFORMER
SECONDARY WINDING

L

N

EARTHED
APPARATUS

EARTH RETURN PATH

Fig. 13.3. Earth loop impedance path.

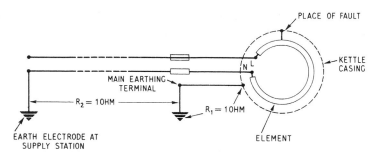

quarter along the length of its element from the 'line' terminal, as shown in the diagram. It is protected by a 15 A fuse which is designed to blow at a current of 47 A in one second, 33 A in ten seconds, or 28 A in one minute.

Assuming that the resistance (R_1) of the earth continuity path between the kettle and the main earthing terminal provided by the supply authority is 1 Ω, and the resistance (R_2) between the earthing terminal and the earth electrode at the supply station is also 1 Ω, ignoring the resistance of the circuit wiring and also that between the element and the point of fault contact with the earth casing, calculate:

(a) the full current which will flow,

(b) the voltage between the metal casing of the kettle and the earth electrode at the supply station, and further,

(c) state which, in your opinion, would burn out, the fuse or the element?　　　　　　　　　　　　　　　　　　　　　　　　　　　　[C]

(a) Redrawing as in Fig. 13.4(a), the diagram clearly shows that the conditions produce a series–parallel circuit. *I* is the total

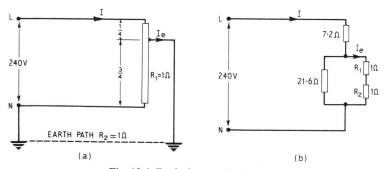

Fig. 13.4. Equivalent earth circuits.

current drawn from the mains, I_e being the fault current flowing through the earth.

$$\text{Resistance of the element} = \frac{V^2}{P}$$

where V is the supply voltage and P the wattage of the element

$$= \frac{240^2}{2000} \qquad = 28 \cdot 8 \, \Omega$$

$$\text{Resistance of } \tfrac{1}{4} \text{ element} = \frac{28 \cdot 8}{4} \qquad = 7 \cdot 2 \, \Omega$$

$$\tfrac{3}{4} \text{ element resistance} = \frac{28 \cdot 8 \times 3}{4} \qquad = 21 \cdot 6 \, \Omega$$

The resistance values may now be incorporated into an equivalent circuit (Fig. 13.4(b)). Resistance of parallel section (R_p) is given by:

$$\frac{1}{R_p} = \frac{1}{21 \cdot 6} + \frac{1}{R_1 + R_2} = \frac{2 + 21 \cdot 6}{43 \cdot 2} = \frac{23 \cdot 6}{43 \cdot 2}$$

$$R_p = \frac{43 \cdot 2}{23 \cdot 6} \qquad = 1 \cdot 83 \, \Omega$$

$$\text{Total resistance} = 7 \cdot 2 + 1 \cdot 83 \qquad = 9 \cdot 03 \, \Omega$$

$$\text{Total current from supply } (I) = \frac{240}{9 \cdot 03} \qquad = 26 \cdot 6 \, A$$

The current through the faulty section (I_e) may be calculated by the inverse proportion resistance method

$$I_1 = I \times \frac{R_2}{R_1 + R_2}$$

$$\therefore \qquad I_e = 26 \cdot 6 \times \frac{21 \cdot 6}{21 \cdot 6 + 2} \qquad = \underline{24 \cdot 4 \, A}$$

(b) Voltage between kettle body and earth electrode is calculated by:

$$I_e \times \text{resistance of earth loop} = 24 \cdot 4 \times 2 \qquad = \underline{48 \cdot 8 \, V}$$

(c) Kettle element is rated at 2 kW at 240 V. This gives a current of

$$\frac{P}{V} = \frac{2000 \, W}{240 \, V} \qquad = \underline{8 \cdot 32 \, A}$$

The current flowing through the shorter portion of the element when in a fault condition is 26·6 A. The excess current would be likely to burn out the element before the fuse.

Comment. This is a wordy question which requires careful thought before commencing an answer. As in many types of problems, to assist in visualising the actual conditions a simple diagram is essential in the first instance. Such a diagram will help to focus the student's thoughts towards finding the correct solution.

The Electricity Supply Regulations, 1937 states that 'connected with earth' means connected with the general mass of earth in such a manner as will ensure at all times an immediate and safe discharge of energy.

13.4. Earthing Practice

Throughout every installation operating above extra-low voltage a circuit protective conductor (c.p.c.) must be provided. An understanding of the term, circuit protective conductor (c.p.c.) is so important that the I.E.E. rendering is here repeated in full:

Circuit protective conductor. A protective conductor connecting exposed conducting parts to the main earth terminal.

The c.p.c. formerly known as the 'earth-continuity conductor' may consist of metal conduits, trunking or ducting. In addition to the protective (earth) wire in p.v.c. sheathed cables and flexible cords or cable metal sheath or wire armouring.

All installation metalwork must be connected to the c.p.c., which in turn is connected to a consumer's earthing terminal. This consumer's earthing terminal shall be provided adjacent to the consumer's supply terminals. The final sub-circuit c.p.c. is required to be connected at an earthing terminal at every lighting point and switch position. For socket outlets mounted in earthed metal boxes a connexion from the earth terminal in the box to the corresponding socket earth terminal is required, in certain circumstances, to be made by a conductor of minimum size 2·5 mm² when associated with a conductor of 2·5 mm².

This is a general part of the demand as stated in I.E.E. Table 54F which requires that, *up to* 16 *mm², the cross-sectional area of the c.p.c. must not be less than its associated phase conductor unless calculations made by* $a = \dfrac{\sqrt{I^2 t}}{k}$

Unless protected by cable sheathing, an e.c.c. must now be protected by insulation at least to single-cored non-sheathed cable of appropriate size complying with BS 2004.

With certain exceptions, e.g. shaver socket conforming to BS 3052, one point of the secondary winding of transformers must be earthed. In a bell transformer, for example, there is a great danger of mains voltage appearing at the secondary under fault conditions. Here the c.p.c. is usually connected to the centre terminal of the common 3-V, 5-V and 8-V transformer.

13.5. Bonding

The term 'bonding' is often confused with earthing. Bonding applies to the internal installation and is in the form of a low-resistance connexion between any two points of the earthed system to prevent a *difference of potential* arising between the points to be bonded. A current cannot be set up without a p.d., therefore bonding can be an effective means of preventing the emergence of earth-leakage currents. The consumer's earth terminal must be properly bonded to the gas and water services as near as possible to the point of entry into the premises (Fig. 13.5). Before bonding connexion is made this main earth point must be suitably tested (see Chapter 14). 'Earthing clips' made of thin copper strips with punched holes are to be avoided; only substantial clamps to BS 951 are permitted.

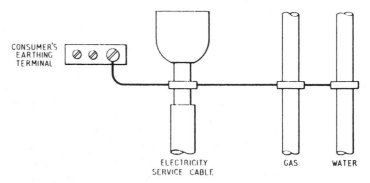

CONSUMER'S EARTHING TERMINAL

ELECTRICITY SERVICE CABLE   GAS   WATER

Fig. 13.5. Cross-bonding of all services is now required and the earthing lead must be brought to the consumer's earth terminal.

Bonding leads must be at least half the size of earthing leads with a .minimum of 6 mm². Connexion by the circuit protective conductors to the supply undertaking's earth terminal, creates an *equipotential zone* within the premises.

Bonding must also be carried out between earthed metal and extraneous metalwork which might fortuitously (accidently) come in contact. The extraneous metalwork includes:

(*i*) Bath and exposed metal pipes, radiators, sinks and tanks, in the absence of metal-to-metal joints of negligible electrical resistance.

(*ii*) Accessible structural steelwork, wherever this is practical.

(*iii*) Framework of mobile equipment on which electrical apparatus is mounted, such as cranes and lifts.

Where bonding under such circumstances is not possible then there must be effectual segregation between these various metalworks to prevent the possibility of appreciable voltage differences.

13.6. Bathrooms

Due to the considerable risk of severe electric shock, special requirements are necessary. Only skirted lampholders are allowed although totally enclosed lighting fittings are recommended. Switches or other means of control must be inaccessible to a person using the bath; the switch should be of the cord-operated type or situated just outside the room entrance door.

Socket-outlets are not permitted in bathrooms (with the exception as described below), neither should any other provision be made for connecting portable apparatus. No stationary appliance having heating elements that can be touched is to be installed within reach of a person using the bath or shower.

Shaver sockets complying to BS 3052 are the only type of socket outlet that may be fitted in a bathroom. This type of socket incorporates a double-wound transformer and while the shaver earth terminal must be connected to its appropriate e.c.c., *the secondary winding of the transformer must not be earthed.* Students are often puzzled by these seemingly contradictory requirements. The earth terminal is connected to the transformer core so that the excess-current protective device would operate in the event of the primary insulation breaking down. The unit forms a 1·1 double wound transformer so that the secondary is completely isolated from the supply. If a person were to come in contact with a live secondary conductor and earth there would be no return path to the neutral at the sub-station (refer to Fig. 13.3), thereby forming a safeguard against the possibility of an electric shock. However, this should not be deliberately attempted.

13.7. Earth Electrodes

Three methods may be employed for earthing, none of which includes underground water pipes:

(1) The supply authority's earth point connected to the consumer's earthing terminal (Fig. 13.5). This poses no problems for the consumer or installation engineer and is to be preferred, the Electricity Board's service-cable sheath providing a low-impedance metallic return path. The earth return may, in country areas, consist of an overhead continuous earth wire.

(2) Where a means of earthing has not been supplied by the supply undertaking, connexion to the consumer's earth terminal may be by an earth electrode consisting of copper rods or strips. The buried copper or galvanised earth plate has gone out of favour, usually being found unsatisfactory by modern standards.

Most of the resistance of these electrodes to earth lies on the top soil, so that driven rods are effective in reducing this resistance. To withstand hammer blows the rods may be copper-clad steel-cored. The rods are threaded for the connexion of successive lengths, which may be up to about 3 m, a common diameter being 16 mm. The actual depth to which the rods may be driven is determined by the type of soil. Typical resistivity values are given in Table 13.1.

TABLE 13.1. SOIL RESISTIVITY

Composition of soil	Resistivity (ohms/cm³)
Clay	2000–6000
Sandy clay	8000–20 000
Marsh, peat	15 000–30 000
Sand	25 000–50 000
Rock	up to 1 000 000

Soil-resistivity values may vary with temperature and season, deeper rods producing more stable conditions. To reduce top soil resistance, the soil may be treated with common rock salt, copper sulphate or magnesium sulphate.

A typical terminal connexion is shown in Fig. 13.6.

For rocky soil, copper strips may be attempted. To produce minimum resistance they may be laid in shallow trenches and connected in the form of a grid.

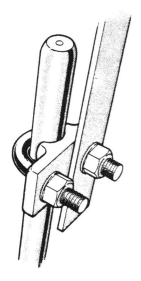

Fig. 13.6. Driven rod terminal connexion.

(3) A special system of earthing known as protective multiple earthing (p.m.e.) may be employed (Fig. 13.7). The use of p.m.e. is on the increase in country areas due to the vagaries of earth resistance. It will be noted that the neutral distribution line is connected at intervals (which may be at every third pole) to earth electrodes. With the system, earth-fault current will flow back to the source of supply by means of the neutral in parallel with the earth, so ensuring that a low resistance path will enable the fuse to blow.

The use of the system must be with the approval of the Department of Energy and the Post Office, as specified in the Electricity Supply Regulations, 1937. Supply undertakings concerned are to be consulted as to any special requirements concerning the size of protective conductors.

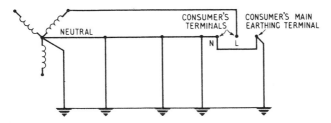

Fig. 13.7. Protective multiple earthing.

13.8. Earthing Conductors and Circuit Protective Conductors

The earthing lead is the final conductor to the earth electrode. The earthing conductor must have a minimum size of 16 mm² with protection against corrosion and 25 mm² if of copper (50 mm² if of steel) where such protection is not offered. Aluminium and copper-clad aluminium conductors are not permitted for final connexions to earth electrodes.

Calculation of the protective conductor size can be obtained by using the I.E.E. adiabatic equation

$$a = \frac{\sqrt{I^2 t}}{k}\,\text{mm}^2$$

I = fault current (A)
t = operating time of protective device (s)
k = values as given in Section 13.2

Essential requirements for sound earthing are that all earthing joints must be tight, mechanically sound and protected against mechanical damage and corrosion—a particular hazard being at the actual point of connexion to the earth electrode. A label is required to be permanently attached at this point with the words 'SAFETY ELECTRICAL EARTH—DO NOT REMOVE', in a legible type not less than 5 mm high. *A separate c.p.c. is necessary for flexible conduit.*

13.9. Voltage-operated Earth-leakage Circuit-breaker (E.L.C.B.)

The general principle of the voltage-operated e.l.c.b. may be seen from the basic circuit shown in Fig. 13.8. This type of circuit breaker is constructed to BS 842, which permits a maximum voltage rise at the earthed metalwork of 40 V. It will operate on a high loop imped-ance. On one tested the device tripped at a loop impedance of 900 Ω.

It will be noticed that the trip coil, being in the earth circuit, is operated only by earth-leakage currents. The weakness in this kind of circuit-breaker lies in the possibility of *parallel earth paths that may render the trip coil inoperable.* For instance, connexion between the circuit protective conductor and the water-pipe system of an

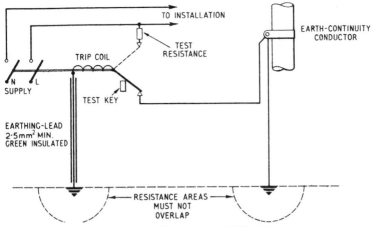

Fig. 13.8. Voltage-operated E.L.C.B.

immersion heater has often provided a low resistance shunt which shorts out the trip coil.

13.10. Residual Current Circuit-Breaker

The residual current circuit-breaker (r.c.c.b.) is often referred to as the core-balance type. A much simplified diagram is given in Fig. 13.9. Here, the live and neutral conductors pass through or are wound on a frame similar to a transformer core, although the frame is usually circular or toroidal in order to form to provide a strong magnetic circuit. The action is based on the fact that for healthy circuits the currents in the line and neutral are the same value so that

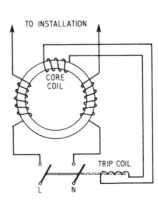

Fig. 13.9. Residual current circuit-breaker.

the magnetic fluxes from these conductors are equal. Thus the fluxes in a normal circuit are balanced. Should an earth fault occur more current flows in one conductor than the other so that an unbalance occurs. The changing flux generates an e.m.f. in the core coil and the trip coil operates. The circuit-breaker may also act as a main switch.

Again, the maximum voltage rise at earthed metalwork is limited to 40 V and it is clear that there is no parallel earth problem associated with the current-operated unit. The operating, or out-of-balance, current must not exceed 2% of the normal current, the value being 1·2 A on a 60 A circuit. Highly-sensitive c.c.b.s, which operate at as low as 6 mA, are available.

Overload current protection may be incorporated in the one unit and they can also be fitted to work on a balanced 3-phase supply.

13.11. Monitored Earth Circuits

With earthing as a means of protection, a break in the earth-continuity conductor may not be noticed until a fault occurs. This condition may be particularly dangerous when handling

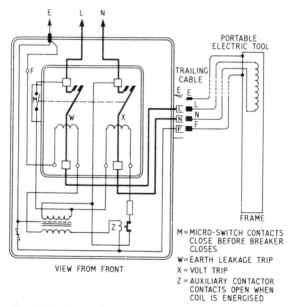

Fig. 13.10. Connexions for single-phase, automatic earth proving supply point.

portable tools in hazardous conditions. An earth-proving unit of the type shown in Fig. 13.10 avoids such dangers. The transformer secondary from the unit passes a small current continuously through two earth-continuity conductors which are connected to the casing of the tool (or other apparatus). If a break occurs, this monitoring circuit is interrupted and a relay cuts off the supply.

EXERCISES

1. (a) Discuss the procedure necessary to measure the resistance of an earth electrode.

(b) Describe one application to afford satisfactory protection against excessive leakage for an installation where water pipes are not available and the supply is by overhead cable. [C]

2. (a) Describe four methods that may be employed to reduce electrode soil resistance without increasing the number or surface area of the earth electrodes.

(b) State why and give the circumstances in which one of the methods is not suitable. [C]

3. (a) What is the purpose of 'earth loop' testing?

(b) Show by diagram the whole of the loop included in such a test.

(c) Outline any two conditions which might lead to high test figures and explain why.

4. Regulations require that 'apparatus operating at a voltage exceeding extra low voltage shall be protected against dangerous earth leakage by one of four methods'.

(a) Give any two of these methods and discuss means of compliance.

(b) State any five of the exemptions to the requirement for earthing metal associated with an installation. [C]

5. Draw the circuit diagrams and compare the relative merits of voltage-operated and current-balance earth-leakage trips.

6. (a) What are the advantages of using an earth-loop impedance test over a continuity test taken between an outlet and the consumer's earth terminal using a standard continuity tester?

(b) Draw the full circuit diagram of the loop used in a line-earth loop impedance test.

(c) Discuss any two conditions which may lead to a high test reading. What measures should be taken to correct the fault in each case? [C]

7. Sketch the complete internal wiring for a typical installation as connected to a p.m.e. system of supply. The wiring to be carried out with mineral insulated cables.

Chapter 14

TESTING

All installations must be thoroughly tested before they are put into commission. The I.E.E. Regulations lay down a sequence of tests which includes proper inspection by an experienced eye. The tests are applicable to new work and apply also to major alterations and extensions. Such tests and inspections should be made at regular intervals.

14.1. Earth-continuity Conductors

Continuity tests on the earth continuity conductors of a.c. installations may be made with a standard supply connected to a transformer, or with either a.c. or d.c. hand testers.

If the secondary of a transformer is used for carrying out the actual tests, it should supply a current approaching 1·5 times the rating of the final sub-circuit, but the maximum value of test current (I) need not exceed 25 A and the voltage (V) applied should not exceed 50 V.

14.2. Earth-electrode Resistance

Due to an insufficiently-low earth-loop impedance, it may be necessary to check the actual resistance of the earth electrode to earth. The measuring procedure is set out in Appendix 6, Section 4, of the I.E.E. Regulations. In general, the resistance of an electrode to the general mass of earth depends upon

(a) shape and material of electrode,

(b) depth of soil in which buried,

(c) resistivity of surrounding soil, which may be seasonal and in turn dependent upon moisture content.

Whenever current passes through an earth electrode there is a *resistance area*, which may be defined as the area where a 'voltage gradient' exists. The term voltage gradient may be understood by the graph (Fig. 14.1), obtained by plotting the voltage between the electrode and earth against distance from the electrode. The voltages

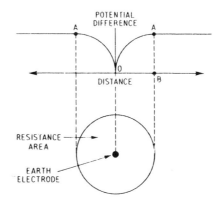

Fig. 14.1. Potential gradient surrounding earth electrode.

are obtained by the product of the resistance (at different points from
the electrode) and the current passing through the soil. The potential
gradient, which is given by 0–A in the graph, is seen to increase
rapidly at first, becomes less and then falls to zero at the point A,

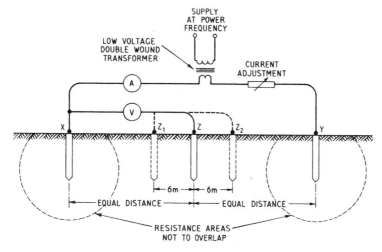

Fig. 14.2. Circuit for measurement of earth-electrode resistance.

X = Earth electrode under test and disconnected from all other sources of
    supply.
Y = Auxiliary earth electrode.
Z = Second auxiliary earth electrode.
Z_1 = Alternative position of Z for check measurement.
Z_2 = Further alternative position of Z for check measurement.
V = High reading (200 Ω/V) voltmeter.

from which the resistance of the mass of the earth does not appreciably change with distance along the top soil from the electrode.

Referring to Fig. 14.2, the resistance is obtained by a steady alternating current between the earth electrode X and an auxiliary earth electrode Y placed at such a distance from X that the resistance areas of the two electrodes do not overlap. A second auxiliary electrode Z is then inserted between X and Y and the voltage drop between X and Y measured.

The resistance of the earth electrode is then the

$$\frac{\text{voltage between X and Z}}{\text{current between X and Y}}$$

provided there is no overlap between the resistance areas. As a check, two further readings must be taken with the second electrode Z moved to the positions Z_1 and Z_2. If the three results are substantially in agreement, the average of the three readings may be taken as the resistance of the electrode X. To obtain agreed results the distance between X and Y may have to be increased. The electrode resistance is thus seen as the maximum potential gradient A–B (Fig. 14.7) divided by the current.

One precaution to be adopted is that the electrode X must be disconnected from all sources of supply other than that used for testing. The Megger Earth Tester has the advantage over the voltmeter–ammeter method as it is not affected by stray currents in the soil; these currents may be return currents from electrical traction sources or leakage currents from distribution systems.

14.3. Insulation Tests

A poor insulator has a comparatively low insulation value and *vice versa*. For this reason, cable and apparatus insulation resistance is expressed in megohms ($10^6 \, \Omega$). Dampness is a major cause of low insulation. The moisture, which is absorbed by the insulation, produces in cables leakage currents which flow outwards *through* the insulation and may be taken to form parallel paths (Fig. 14.3). Thus,

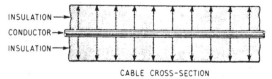

INSULATION →
CONDUCTOR →
INSULATION →

CABLE CROSS-SECTION

Fig. 14.3. Leakage current, flowing outwards through cable resistance form parallel paths.

insulation resistance is inversely proportional to the cable length. Similarly, the insulation resistance of several circuits will be less than that of any one of the circuits.

14.4. Verification of Polarity

Checking that all single-pole switches, thermostats and other control devices are in the live side is a statutory requirement of the E.S.R. 27 (i) (a). The polarity test is also made to ensure that the outer contacts of centre-contact bayonet and Edison-type screw lamp-holders are connected to the neutral, and that wiring has been correctly connected to plugs and socket-outlets. Crossed polarity, sometimes termed 'switched neutrals', is one of the major causes of severe electrical accidents.

A typical report from a Home Office bulletin reads as follows: 'A 71-year-old man received a fatal shock from an electric bowl fire. The fire was switched off but the wall socket was incorrectly wired, giving reversed polarity. Four of the other five sockets in the house were similarly wired.'

For new work, where there is no supply, a bell or continuity set (Figure 14.4) will serve the purpose; for switched socket-outlets a purpose made tester can be used to check that the switch is in the phase conductor. In addition, circuit and initial continuity tests may be carried out with the simple test set, but as it does not measure resistance it is not suitable for checking the conductivity of the earth-continuity conductors when testing the effectiveness of earthing.

Neon test screwdrivers are commonly employed where the mains supply is available but care is required. A properly guarded pigmy filament lamp with h.b.c. fuse and shrouded prods has many advantages for this purpose.

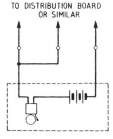

TO DISTRIBUTION BOARD
OR SIMILAR

Fig. 14.4. Connexions at mains position for polarity and continuity testing.

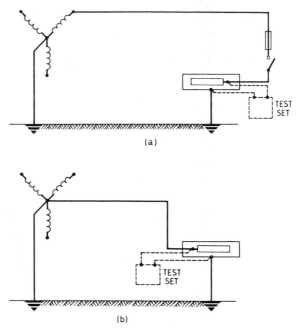

Fig. 14.5. (a) Line-earth loop path. (b) Neutral-earth loop path.

14.5. Loop Impedance Tests

From Fig. 13.3 it can be seen that there are two loop impedance paths, namely line-earth and neutral-earth (Fig. 14.5), either of which may be tested by appropriate test sets in accordance with Appendix 15, item 5 of the I.E.E. Regulations. The loop paths are also illustrated in Appendix 6 of the Regulations.

It must be borne in mind that the purpose in both methods of test is to ensure that the loop impedance is low enough to allow sufficient current, in the event of an earth fault, to operate the protective gear. According to the type of instrument, the reading may be:

(1) a direct reading in ohms, or

(2) a direct indication of the maximum permissible fuse rating or setting of a circuit-breaker, or

(3) an indication whether a fuse or circuit-breaker of a given current rating will operate satisfactorily under fault conditions, often termed 'go' or 'no-go'.

The full mains voltage is employed for the line-earth type of test, which in one type passes a short-duration (0·25 second) current of 20 A through a 10 Ω resistor in the tester. For this reason, care

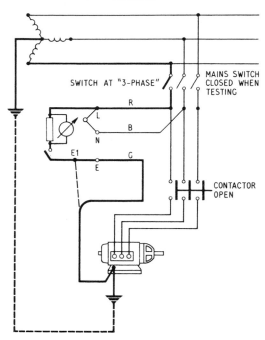

Fig. 14.6. Testing the earthing of 3-phase apparatus
where no neutral is available.

must be taken that no ill effects can arise if the earthing is defective.
In this respect, a particular source of trouble can occur if the wiring
runs adjacent to gas-pipes. Fig. 14.6 shows the line-earth loop test
where no neutral is available.

Neutral-earth loop testers, which must not be used on protective
multiple earthing systems, operate by injecting a supply into this
loop, the current is obtained from the secondary of a transformer;
the unit may incorporate a voltmeter and ammeter (Fig. 14.7).

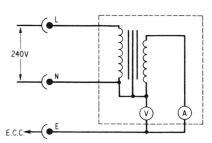

Fig. 14.7. Essentials of neutral-earth
loop impedance tester.

Alternatively, a rapidly-reversing d.c. source is permitted, provided that any inductors in the circuit protective conductor are disconnected at the time of the test. Since currents may be flowing in the neutral from other consumers, instruments are designed so that the indications are not affected by these currents. Measurements on systems fed from small transformers may require compensation to include an allowance for the impedance of the windings of transformers, balancers, etc.

On screwed-conduit installations, the impedance is likely to be appreciably above the measured resistance value. If measurements are made with less than 10 A a.c. or with a rapidly-reversing d.c. and the e.c.c. is wholly or mainly of steel conduit, the effective value of impedance is to be taken as twice the measured value, less the reading taken at the consumer's earth terminal.

14.6. Earth-leakage Circuit-breakers

Testing requires a voltage not exceeding 45 V obtained from a double-wound transformer connected to mains supply to be applied across the neutral and earth terminals (or the neutral and frame terminals of a fault-voltage operating device). Under these conditions the circuit-breaker is to trip instantaneously. The test-key of an e.l.c.b. should be pressed and the operation checked at intervals not exceeding three months and a maintenance inspection made regularly.

14.7. Measurement of Earth Resistivity

The earth tester may also measure soil resistivity. Four metal spikes are driven into the ground, up to a depth of 1 m and not exceeding one-twentieth of their separation, in a straight line with typical dimensions as given in Fig. 14.8. With homogeneous soil,

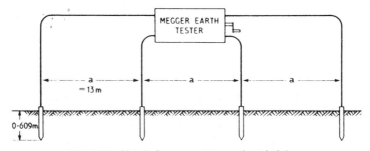

Fig. 14.8. Circuit for measuring earth resistivity.

the average resistance to a depth of $a/20$ is given by $\rho/2\pi a$, where a is the electrode spacing in cm and the resistivity in Ω-cm, provided the resistance areas of the electrode do not overlap. As a point of interest it may be noted that this method is used extensively in preliminary geological surveys and for confirming the presence of mineral deposits.

Example 14.1

Three separate circuits are disconnected from a fuse board and tested for insulation to earth. If the values are 4 MΩ, 6 MΩ and 30 MΩ, what is the combined insulation resistance to earth?

Combined insulation conductance $= \frac{1}{4} + \frac{1}{6} + \frac{1}{30} = \frac{27}{60} = \frac{9}{20}$

Combined insulation resistance $= \frac{20}{9}$ $\quad = \underline{2 \cdot 2 \text{ M}\Omega}$

There is another type of leakage termed *surface leakage*. Being due to dampness, it also provides current-leakage paths and will be increased by any dirt on the insulator. Surface leakage takes place between insulated conductors and earth, or between conductors of opposite polarity.

For insulation tests of large installations, the outlets (an outlet consists of a point or switch; a switch-socket or switched appliance also counts as one outlet) may be divided into groups of not less than 50. Insulation measurements are normally made by a d.c. voltage of 500 V supplied by a test instrument which also incorporates a high-reading ohmmeter. The I.E.E. Regulations require three tests:

(1) TO EARTH. With all fuse-links in place, switches closed, all poles and phases connected together (except in earthed concentric wiring) the insulation resistance to be not less than 1 MΩ.

(2) BETWEEN POLES OR PHASES. If possible all lamps to be removed, apparatus disconnected and switches closed. The insulation resistance again not to be less than 1 MΩ. This test is not applicable to earth concentric wiring.

(3) APPARATUS. The insulation resistance between all live parts and the case or frame to be not less than 0·5 MΩ to earth and between poles or phases.

Every ring circuit must be tested to verify the continuity of all conductors, including the earth-continuity conductor. This is usually carried out at the 30 A fuseway.

14.8. Certificates

After the testing and inspection, the I.E.E. Completion and Inspection Certificates should be given by the contractor or other authorised person, who should also remind the consumer of the importance of periodic testing and inspection.

The following durable notice in indelible letters not smaller than 12-point should be fixed at the main intake of every installation upon completion of the work:

IMPORTANT

THIS INSTALLATION SHOULD BE PERIODICALLY INSPECTED AND TESTED, AND A REPORT OF ITS CONDITION OBTAINED, AS PRESCRIBED IN THE REGULATIONS FOR THE ELECTRICAL EQUIPMENT OF BUILDINGS ISSUED BY THE INSTITUTION OF ELECTRICAL ENGINEERS.

14.9. Statutory Requirements

By virtue of the Electricity Supply Regulation No. 6, the Area Boards are empowered to refuse connexion to an installation where the leakage current exceeds one ten-thousandth part of the maximum current supplied to the installation. From this requirement some values have been calculated as shown in Table 4.

It will be seen that the I.E.E. Regulations are much more stringent in this respect than the E.S.R.

14.10. Measurement of Power and Power Factor

For measuring power and power factor the usual method is to fit an ammeter, voltmeter and wattmeter in the circuit, from which

TABLE 4. STATUTORY INSULATION REQUIREMENTS

Maximum current supplied	Maximum leakage current in mA	Minimum insulation resistance in MΩ (240 V supply)
100 A	10	0·024
60 A	6	0·04
30 A	3	0·08
20 A	2	0·12

the power factor, $\cos \phi = P/VI$ can be calculated, P being the power as read by the wattmeter.

This method can also be adopted for a sinusoidal 3-phase balanced load provided the star point is available. The wattmeter current coil is placed in series with one line and the voltage coil is connected between any one line and the star point, the total power being three times the power as read by the single-phase wattmeter. Where the star point is not readily obtainable, the two-wattmeter method forms a convenient means of obtaining power and power factor.

14.11. Two-wattmeter Method

Figure 14.9 shows the connexions to the stator of a 3-phase induction motor. Currents I_R, I_Y and I_B show positive directions.

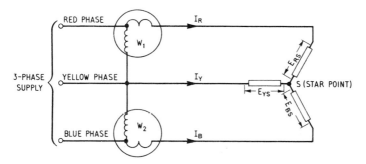

Fig. 14.9. Two-wattmeter connexions.

P.D. across voltage coil of W_1, E_{RSY} = phasor difference between
$$E_{RS} \text{ and } E_{YS}$$
$$= E_{RS} - E_{YS}$$
P.D. across voltage coil of W_2, $E_{BSY} = E_{BS} - E_{YS}$

These values may now be shown in the phasor diagram (Fig. 14.10).

The phase difference between E_{RSY} and $I_R = 30 + \phi$

∴ Reading on W_1, $P_1 = E_{RSY} I_R \cos(30° + \phi)$

The phase difference between E_{BSY} and $I_B = 30° - \phi$

∴ Reading on W_2, $P_2 = E_{BSY} I_B \cos(\phi - 30)$

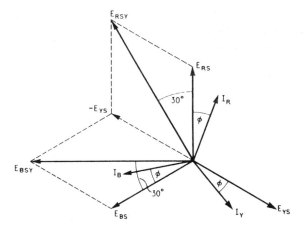

Fig. 14.10. Phasor diagram for 2-wattmeter method.

Since the motor windings are balanced

$$P_1 = EI \cos(30° + \phi)$$

$$P_2 = EI \cos(30° - \phi) \quad \text{where} \quad I = I_R = I_Y = I_B$$

$$E = F_{RSY} \quad = E_{BSY}$$

$\therefore$ Total power $P_1 + P_2 = EI \cos(\phi + 30) + EI \cos(\phi - 30)$

Collecting terms and expanding the trigonometrical expressions,

$$P_1 + P_2 = EI[(\cos 30° \cos \phi - \sin 30° \sin \phi) + (\cos 30° \cos \phi$$
$$+ \sin 30° \sin \phi)]$$
$$= EI \times 2 \cos 30° \cos \phi \qquad\qquad (\cos 30° = \sqrt{3}/2)$$
$$= EI \times 2 \times \sqrt{3}/2 \cos \phi$$
$$= EI \sqrt{3} \cos \phi$$

$$P_1 - P_2 = EI[\cos(30° + \phi) - \cos(30° - \phi)]$$

$$= EI[(\cos 30 \cos \phi - \sin 30 \sin \phi) - (\cos 30 \cos \phi$$
$$+ \sin 30 \sin \phi)]$$

$$= EI(- \sin \phi)$$

$$\therefore \quad P_2 - P_1 = EI \sin \phi$$

$$\frac{P_2 - P_1}{P_1 + P_2} = \frac{EI \sin \phi}{\sqrt{3} \, EI \cos \phi} \qquad \left(\frac{\sin \phi}{\cos \phi} = \tan \phi \right)$$

$$\therefore \quad \tan \phi = \sqrt{3} \times \frac{P_2 - P_1}{P_1 + P_2}$$

From which the power factor, $\cos \phi$ may be obtained. With reference to the phasor diagram (Fig. 14.10), when the angle of lag is 60° the wattmeter measures

$$VI \cos(30° + 60°) = 0 \qquad \text{since } \cos 90° \text{ is zero}$$

So that when the power factor drops to $\cos 60°$ (0·5), W_1 reads zero. With a lag of more than 0·5 the meter reads backwards and either the current or voltage coil must be reversed. This forward reading must be subtracted from W_2 to obtain the total power. It can also be seen that when the angle of lead is 60° the second wattmeter reads zero. For power factors less than 0·5, with the current leading, W_2 gives a negative reading and commences to read backwards. Again the current or voltage coil must be reversed and the positive reading subtracted.

Example 14.2

The steady readings obtained from two single-phase wattmeters used to measure the power supply to a three-phase load are 14 W and 35 kW.

Calculate :

(i) *the total power in kW taken by the load,*
(ii) *the power factor,*
(iii) *the load kVA,*
(iv) *the line current, if the supply is 415 V, three-phase.*

(i) Total power $= P_1 + P_2 = 0·014 + 35 \qquad = \underline{35·014 \, \text{kW}}$

(ii) $\qquad \tan \phi = (\sqrt{3}) \left(\frac{P_2 - P_1}{P_1 + P_2} \right)$

$$= \sqrt{3} \times \left(\frac{35 - 0·014}{0·014 + 35} \right) = \sqrt{3} \times \left(\frac{34·986}{35·014} \right) = 1·731$$

$$\phi = 60°$$

$$\cos \phi = \cos 60° \qquad\qquad\qquad = \underline{0·5}$$

(*iii*) $kW = kVA \cos \phi$ $= 70{\cdot}03 \, kVA$

∴ $kVA = \dfrac{35{\cdot}014}{0{\cdot}5}$

(*iv*) $Power = \sqrt{3} \, VI \cos \phi$

$35{\cdot}014 \times 1000 = \sqrt{3} \times 415 \times I \times 0{\cdot}5$

∴   Line current $I = \dfrac{35{\cdot}014 \times 1000}{\sqrt{3} \times 415 \times 0{\cdot}5}$ $= 97{\cdot}4 \, A$

14.12. Murray Loop Test

The test is an application of the Wheatstone Bridge and is used for the location of earth faults in underground cables. Fig. 14.11 shows one arrangement.

R_1 and R_2 are adjusted until zero reading is obtained on the galvo, so that it balances

$$\frac{R_1}{R_2} = \frac{2l-x}{x}$$

$$xR_1 = 2R_2 l - R_2 x$$

$$xR_1 + R_2 x = 2R_2 l$$

$$x(R_1 + R_2) = 2R_2 l$$

$$x = \frac{2lR_2}{R_1 + R_2}$$

from which x can be obtained.

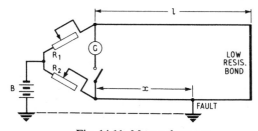

Fig. 14.11. Murray loop test.

$R_1 R_2$ = Variable resistors.
B = Battery.
G = Galvanometer.
l = Length of sound core.
x = Distance of fault from test set.

Example 14.3

In a Murray loop test a balance was obtained with $R_1 = 90 \, \Omega$ and $R_2 = 10 \, \Omega$ on a faulty 2-core cable of length 150 m. Find the distance of the fault from the test end.

$$x = \frac{2lR_2}{R_1 + R_2} = \frac{2 \times 150 \times 10}{90 + 10} = 30$$

Hence fault lies at a distance of 30 m from the test end

In practice the test would be repeated from the other end and the average of both calculations taken. The variable resistors may be replaced by a slide wire bridge. Since it is a question of measuring a low resistance, Bridge Meggers are often employed for this purpose. The above test requires one sound core. Where both cores are down to earth, an auxiliary wire must be used.

For breaks or open-circuits, capacitor bridges are employed. Both kinds of faults may also be found by proprietary tests which employ electronic means for the location of the fault position.

Whatever method is adopted a logical approach is necessary. Records of the cable runs render assistance in fault finding.

EXERCISES

1. Describe the instruments required and the methods you would adopt to make the following tests on an installation: (*a*) the insulation resistance of conductors, (*b*) continuity of conduits.

Give in each case the basic figure which would satisfy I.E.E. Regulations.
[C]

2. Explain the principle of the simple 'slide-wire potentiometer'.

Describe, with diagrams, the use of a direct-reading potentiometer to calibrate: (*a*) a d.c. ammeter, 0–20 A, (*b*) a d.c. voltmeter, 0–250 V. [C]

3. (*a*) What is meant by 'resistance area'?

(*b*) Describe the procedure for measuring the resistance of an earth electrode.

4. A moving-coil galvanometer of resistance 5 Ω, requires a potential difference of 75 mV to give full-scale deflection. Calculate:

(*a*) the value of the shunt resistance needed to enable the instrument to work as an ammeter and to give full-scale deflection at 30 A.

(*b*) the value of the series resistance to allow the instrument to work as a voltmeter with a full-scale reading of 250 V.

With the aid of sketches, explain the action of this instrument. [C]

5. A three-phase p.i.l.c. armoured cable running through a works is found to be faulty. Tests show that one core has a very low resistance to earth. The other cores have satisfactory insulation resistance, and all cores have good continuity.

Explain how you would proceed to locate the position of the fault. Give a diagram of connections of your fault-finding arrangement, and describe the instruments you would use.

Describe briefly how you would make the repair.   [C]

6. Describe briefly, with the aid of clear sketches, the action of BOTH the following: (a) a moving-iron ammeter, and (b) a moving-coil ammeter.

State for each instrument, whether it can be used to measure d.c. or a.c. or both.   [C]

7. (a) Explain in detail the meaning of 'equipotential zone' and its effect on safety.

(b) Describe the action of a wattmeter as used on both a.c. and d.c.

Chapter 15

ILLUMINATION

15.1. Wave Nature of Light

Light is a form of energy which is radiated or sent out from a source in a wave form and is part of a whole family of electro-magnetic waves, wavelength being the distance between the peaks of the wave of energy. We can think of the wavelength, not the wave, in a similar way to the distance between successive waves on the sea. As we shall see, the question of wavelength is crucial in determining the type of light or in fact whether it expresses itself as light at all.

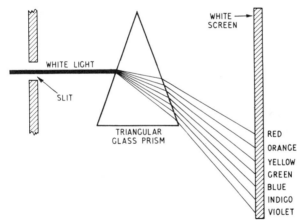

Fig. 15.1. The light spectrum.

The natural light from the sun or a tungsten lamp is often termed 'white light' and consists of a mixture of spectrum of all the colours of the rainbow (Fig. 15.1). From the diagram it can be seen that the rays of light leaving the prism are bent, red the least and violet the most. Figure 15.2 gives the actual wavelength of these colours. Due to the smallness of the wavelength a unit known as a micron is frequency quoted; one micron is equal to one millionth part of a metre, or μm in SI units.

The ultra-violet and infra-red radiations are invisible.

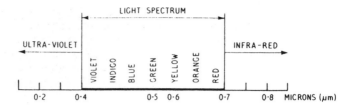

Fig. 15.2. Wavelength in microns (10^{-6} m or μm).

15.2. Illumination Laws

The power of the source of light is known as its LUMINOUS INTENSITY (I). The source was historically the wax candle and is now an agreed standard which is maintained at the National Physical Laboratory. CANDELA (pronounced candeela) is the unit of intensity of a point source which emits light energy from it in all directions. Point source is a relative term as it is assumed to be very small in comparison with the surroundings.

The flow of light or LUMINOUS FLUX (F) which is sent out by the source is measured in LUMENS, one lumen being the light flux emitted within unit solid angle from a point source of 1 candela. Now 1 radian can be considered as the angle subtended by an arc equal to the unit radius r (Fig. 15.3(a)), while a solid angle encloses an area on a sphere equal to the square of the radius (Fig. 15.3(b)).

Since the surface area of a sphere $= 4\pi r^2$
$$= 4\pi \text{ m}^2 \text{ where } r = 1 \text{ m}$$

∴   Number of solid angles in a sphere of 1 m radius $= 4\pi$

Hence 4π lumens are emitted by a point source of 1 candela.

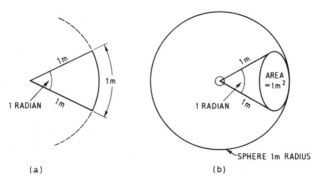

Fig. 15.3. (a) Radian. (b) Solid angle or steradian.

ILLUMINANCE (E) is the light falling on a surface. It is measured in terms of luminous flux *received* on unit area, i.e. *lumens per square metre* (lm/m^2), the unit being the LUX. Illumination follows the inverse square law so that if the surface illuminated is moved twice as far from its original position then the illumination is now a quarter of its former value. For example, if the sphere in Fig. 15.3(*a*) is replaced by one of 2 m radius then

$$\text{Surface area} = 4\pi 2^2 = 16\pi$$

$$\text{Illumination from one candela (cd)} = \frac{\text{lumens}}{\text{area}} = \frac{4\pi}{16\pi} = 0\cdot25\,\text{lux}$$

The three terms are related by the formula:

$$E = \frac{I}{d^2}\,\text{lux}$$

For conversion purposes 1 lumen/ft^2 = 10·76 lux.

Example 15.1

A fitting designed for a shop window gives a light intensity of 1000 cd downwards. Calculate (a) the distance to produce an illumination of 10 lux on a horizontal display counter. (b) If the distance is doubled what must be the power of the source to produce the same illumination?

(*a*) $E = \dfrac{I}{d^2}$ (*b*) $I = Ed^2$

$$d = \sqrt{\frac{I}{E}} = \sqrt{\frac{1000}{10}} \qquad\qquad = 10 \times 20^2$$

$$= \underline{10\,\text{m}} \qquad\qquad\qquad\qquad = \underline{4000\,\text{cd}}$$

PHOTOMETER BENCH. Fig. 15.4 shows a simple laboratory method of comparing the intensities of two lamps. The wedge-shaped

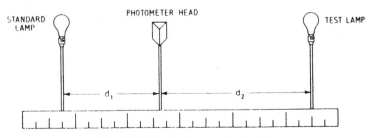

Fig. 15.4. Simple photometer bench.

photometer head has two white matt faces and is adjusted so that when the illumination on the two faces is judged to be equal then

$$\frac{\text{Intensity of standard lamp (cd)}}{d_1^2} = \frac{\text{Intensity of test lamp (cd)}}{d_2^2}$$

Example 15.2

What do you understand by the expression law of inverse squares as applied to illumination problems? Illustrate by means of a sketch.

Two metal filament lamps, with luminous intensities 150 candelas and 300 candelas respectively, are fixed 10 m apart on a level bench. A double-sided matt white screen is placed on the line between the lamps so that each side directly faces one lamp. The screen is positioned so that both sides of the screen are equally illuminated.

Calculate (a) the distance between the screen and the larger lamp, and (b) the illumination on each side of the screen if it were positioned half-way between the lamps.   [C & G]

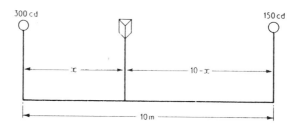

Fig. 15.5. Illustrating inverse-square-law problem.

(a) The facts of the problem may be represented by a sketch as in Fig. 15.5.

Let the distance from the larger lamp to the screen be 10 m, then at balance

$$\frac{300}{x^2} = \frac{150}{(10-x)^2}$$

Cross multiplying $$\frac{(10-x)^2}{x^2} = \frac{150}{300}$$

Taking the square root of both sides $\dfrac{10-x}{x} = \dfrac{1}{\sqrt{2}}$

$$1\cdot414(10-x) = x$$

$$14\cdot14 - 1\cdot414x = x$$

$$2\cdot414x = 14\cdot14$$

$$x = \dfrac{14\cdot14}{2\cdot414} \quad = \underline{5\cdot857 \text{ m}}$$

(b) *When placed half-way,*

Illumination on screen from 300 cd lamp $= \dfrac{300}{5^2} = 12 \text{ lux}$

Illumination on screen from 150 cd lamp $= \dfrac{150}{5^2} = \underline{6 \text{ lux}}$

Manufacturers obtain the total lumen output of lamps by commercial type integrating photometers in the form of a large sphere or cube. This gives the illumination in all directions.

15.3 Cosine Law

We have now to consider the case where the plane to be illuminated is not perpendicular or *normal* to the source. A and B (Fig. 15.6) are two plane surfaces, B is inclined to A by an angle θ. Comparing areas,

$$\dfrac{B}{A} = \dfrac{1}{\cos\theta}$$

Assuming perpendicular rays of light reach A, so that the illumination falling on B (with A removed) is decreased by cos θ, since B presents a larger surface area. Hence the inverse square law is now modified to $E = I \cos\theta/d^2$.

It should be noted that this law is of general application. If there is no inclination, $\theta = 0°$ and cos $0° = 1$.

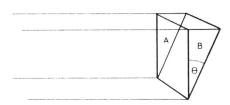

Fig. 15.6. Cosine law.

Example 15.3

A certain incandescent lamp, giving equal illuminance below the horizontal, hangs from the ceiling of a room. The illuminance received on a small horizontal screen, lying on a bench vertically below the lamp, is 63·5 *lux. When the screen is moved horizontally a distance of* 1·3 m *along the bench, the illumination is then* 30·8 *lux.*

Calculate the candle power of the lamp and its vertical distance above the bench. [C]

Referring to Fig. 15.7,

Illumination below the lamp $E_1 = \dfrac{I}{h^2}$

$$I = E_1 h^2 \qquad (1)$$

Illumination 1·3 m along $\qquad E_2 = \dfrac{I \cos \theta}{d^2}$

$$= \frac{Ih}{d^3} \quad \text{since } \cos \theta = h/d$$

$$I = \frac{E_2 d^3}{h} \qquad (2)$$

Equating expressions (1) and (2)

$$E_1 h^2 = \frac{E_2 d^3}{h}$$

$$\frac{h^3}{d^3} = \frac{E_2}{E_1}$$

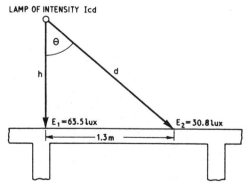

LAMP OF INTENSITY I cd

θ

h

d

$E_1 = 63.5$ lux $E_2 = 30.8$ lux

1.3 m

Fig. 15.7. Illumination along a bench.

But $h/d = \cos\theta$ and taking cube roots of both sides

$$\cos\theta = \sqrt[3]{\frac{30\cdot8}{63\cdot5}} = 0\cdot7856$$

∴ $\theta = 38°\ 14'$

Also $\tan\theta = \dfrac{1\cdot3}{h}$

$$h = \frac{1\cdot3}{\tan 38°\ 14'} = 1\cdot65$$

Vertical distance below the lamp $= \underline{1\cdot65\ \text{m}}$

Substituting in (1) $I = 63\cdot5 \times 1\cdot65^2$ $= \underline{173\ \text{cd}}$

Example 15.4

An advertisement board 4·572 m square is fixed to a wall with the bottom edge near the ground. A lamp fitting giving a luminous intensity of 4000 candelas in all directions towards the board is fixed level with the bottom of the board and at 6·096 m distance, giving a maximum illumination at the centre of the bottom edge. Calculate the illumination:

(a) *at the centre of the bottom edge,*
(b) *at the centre of the top edge,*
(c) *at one of the top corners.*

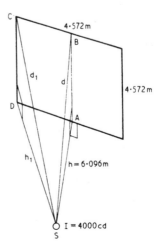

Fig. 15.8. Illumination of advertisement board.

Suggest a method of giving a reasonably even illumination over the whole board. [C as modified]

(a) Illumination at centre of the bottom edge A $= \dfrac{I}{h^2}$

$$= \dfrac{I}{h^2} = \dfrac{4000}{6 \cdot 096^2} \qquad\qquad = \underline{107 \cdot 6 \text{ lux}}$$

(b) From SAB, $d = \sqrt{(6 \cdot 096^2 + 4 \cdot 572^2)}$ $= 7 \cdot 62 \text{ m}$

By Cosine law,

Illumination at B $= \dfrac{Ih}{d^3} = \dfrac{4000 \times 6 \cdot 096}{7 \cdot 62^3}$ $= \underline{55 \cdot 1 \text{ lux}}$

(c) From △ SAD,
$$h_1 = \sqrt{(6 \cdot 096^2 + 2 \cdot 286^2)} \qquad\qquad = 6 \cdot 51 \text{ m}$$
From △ SCD, $d_1 = \sqrt{(6 \cdot 51^2 + 4 \cdot 572^2)}$ $= 7 \cdot 955 \text{ m}$

Illumination at C $= \dfrac{Ih}{d_1^3} = \dfrac{4000 \times 6 \cdot 096}{7 \cdot 955^2}$ $= \underline{48 \cdot 43 \text{ lux}}$

Reasonably even illumination over the board can be obtained by hanging additional angle dispersive-type reflector fittings over the top of the board, or at the two sides.

15.4. Practical Lighting Schemes

The simple inverse-square-law calculations are only really applicable to point sources or where there are no reflecting surfaces, such as may be obtained for outdoor lighting. While interior light fittings produce illumination on the working plane or the surface to be illuminated, there is an additional secondary source of illumination. This is produced by reflection from the fittings themselves, walls and ceilings. Figure 15.9(a) indicates four possible types of reflection which may arise from different building finishes. All fittings may be reduced to five basic types according to their light distribution as shown in Fig. 15.9(b).

Practical schemes are based on the LUMEN METHOD, which takes into account the various factors affecting light flux distribution, among which is the overall light flux likely to be received on the working plane, and also allows for deterioration in the efficiency of the light unit. The lumen method will give a general illumination which might have to be supplemented by local lighting. The modern tendency except where decorative effects are required is to reduce

TYPES OF REFLECTION

DIRECT

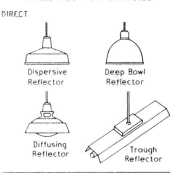

Dispersive Reflector   Deep Bowl Reflector

Diffusing Reflector   Trough Reflector

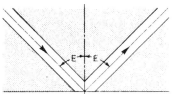

Specular Reflection from a smooth or polished surface: the angle of incidence of any light ray equals the angle of reflection.

SEMI-DIRECT

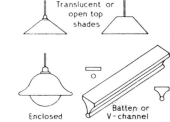

Translucent or open top shades

Enclosed   Batten or V-channel

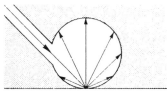

Diffuse reflection from a uniformly matt surface: the brightness of the reflection does not change with change of viewpoint.

GENERAL DIFFUSING

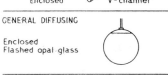

Enclosed Flashed opal glass

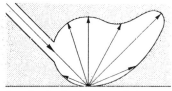

Spread reflection from a semi-glossy surface (egg-shell finish).

SEMI-INDIRECT

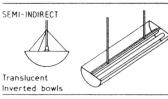

Translucent Inverted bowls

INDIRECT

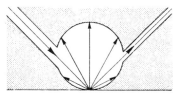

From porcelain or synthetic enamel surfaces, a mixture of diffuse and specular reflection is obtained.

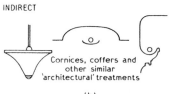

Cornices, coffers and other similar 'architectural' treatments

(a)   (b)

Fig. 15.9. (a) Types of reflection. (b) Lighting fittings, types and performance.

the local units as much as possible by providing very good overall illumination.

Present-day lighting design makes use of the illumination values prepared by the Illuminating Engineering Society (I.E.S.). Catalogues of selected manufacturer's lamps and fittings include *polar curves*, these being special graphs which display the candle power (candelas) in all directions—downwards, upwards and sideways over 360° by means of polar, or circular, co-ordinates. Our calculations are somewhat simplified in order to show the general method although the light specialist may need to account for the many complex details.

15.5. Coefficient of Utilisation

Part of the lumen output from the sources is lost in the fittings. Some output is directed to the walls and ceilings where part will be absorbed and part reflected. Thus, only a portion of the light given out reaches the working surface. This proportion is expressed as a number always less than unity. This utilisation factor clearly shows that the light reaching the plane to be illuminated is reduced so that the power of the light source may have to be increased to achieve the desired value of illumination. The actual determination of the coefficient of utilisation value calls for experience and judgment.

15.6. Maintenance Factor

Dust and dirt on fittings and sometimes ageing of the fittings reduce the light output. Illumination will also be impaired by deterioration in the condition of the decorations. A figure of 0·8 is commonly taken but will have to be reduced for dusty and dirty atmospheres as is sometimes found in woodworking and chemical works. Occasionally the term depreciation factor is used.

$$\text{Depreciation Factor} = \frac{1}{\text{Maintenance factor}}$$

15.7. Spacing-height Ratio

The correct mounting height of fittings is important. Glare may result when they are in the line of vision; excessive height is bound to affect the illumination and make for problems in re-lamping and maintenace. Factories often have the fittings fixed at roof truss or beam height. The value of the spacing-height ratio depends upon the type of fitting and the illumination which may be specified. Once the height is decided, the ratio enables the spacing and therefore the number of fittings to be determined.

15.8. Illumination Formulae

Where total number of lumens are required for a certain building:

$$\text{lumens} = \frac{\text{Illumination (lux)} \times \text{areas of working plane (m}^2)}{\text{Maintenance factor} \times \text{coefficient of utilisation}}$$

Illumination received from a group of light units:

$$\text{lux} = \frac{\text{Total lumens} \times \text{M. factor} \times \text{coeff. of utilisation}}{\text{Area (m}^2)}$$

Example 15.5

A light assembly shop, 15 m long, 9 m wide and 3 m up to trusses, is to be illuminated to a level of 200 lux. The utilisation and maintenance factors are respectively 0·9 and 0·8. Make a scale drawing of the plan of the shop and set out the required lighting points, assuming the use of tungsten lamps and dispersive metallic reflectors. You may assume a lamp efficiency of 13 lm/W, and spacing height ratio of unity. [C as modified]

$$\text{Lumens required} = \frac{E \times A}{M.F. \times C.U.} = \frac{200 \times 15 \times 9}{0·9 \times 0·8} = 37\,500$$

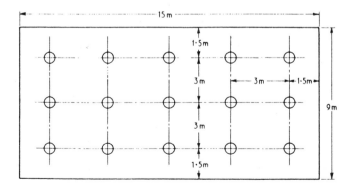

Fig. 15.10. Example 15.5, required lighting points.

With a spacing-height ratio of unity $\dfrac{S}{H} = 1$

and a mounting height of 3 m, spacing $S = 3 \times 1$ $= 3\,\text{m}$

The required lighting points may be set out as in Fig. 15.10.
The layout diagram shows 15 lamp units for even light distribution

With lamp units at 13 lm/W,

$$\text{total power} = \frac{37\,500}{13} = 2885\,\text{W}$$

$\therefore$ power required for each lamp $= \dfrac{2885}{15}$ $= 192{\cdot}3\,\text{W}$

Hence fifteen 200 W lamps are required

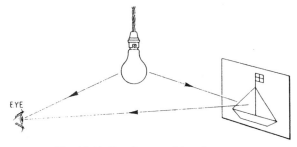

Fig. 15.11. Dazzle caused by glare.

15.9. Glare

Glare may be defined as excessive brightness in the field of vision. In Fig. 15.11 direct light falls on the viewer's eye making it difficult to see the object. There will also be the possible effect of high reflection from the picture's varnish producing an additional source of glare. We require the light to fall on the object and not on the viewer's eye.

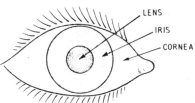

Fig. 15.12. Eye (simplified front view).

To understand why glare inhibits the act of seeing we need to know a little about the action of the human eye (Fig. 15.12). The iris acts as a curtain or shutter to control the amount of light entering the eye. Light as we have seen is a form of radiant energy which passes through the lens to the sensitive nerve layer called the *retina* at the back of the eye. It is then conveyed by the optic nerve to the brain causing the sensation of light. Looking directly at a bright source produces an intense impression on the retina. To avoid damage to this sensitive portion of the eye, the iris automatically contracts, thereby reducing the intensity of the image received. This closing of the iris cuts down the amount of light received. Thus intense bright light in the wrong place actually makes seeing more difficult and also has the effect of producing eye fatigue.

To guard against glare, fittings may now be obtained with illumination details including an I.E.S. Glare Index, enabling calculations to be made which will avoid this harmful effect.

In general, fluorescent fittings with reflectors should be fixed so as to be viewed crossways, whereas bare tubes should lie longways.

15.10. Tungsten Lamp Developments

When a voltage is applied across the ends of a length of wire, the current will be inversely proportional to the resistance of the wire. Reducing the resistance of the wire and therefore increasing the current flowing through it will cause the wire to become red hot and finally white hot, giving off radiation spread over the visible spectrum. It is now heated to incandescence (hence the name 'incandescent' lamp).

Early lamps were made of carbon filaments enclosed in an evacuated glass envelope (Fig. 15.13(a)). Filaments are now constructed of tungsten due to its exceptionally high melting point. At temperatures above 2000°C whilst the filament does not melt it begins to break up and particles fly to the side of the lamp. This causes the glass bulb to become blackened, while causing the weakened filament to show uneven resistance, which sets up hot spots, and the filament burns out and breaks.

Filling the bulb with an inert gas reduces the rate of filament evaporation. To give a higher working temperature with increased efficiency, the cooling effect of convection currents in the gas is reduced by winding the filament into a fine helical coil and then forming into a crescent shape (Fig. 15.13(b)). In single-coil lamps the helix is formed by winding on a steel wire rod or mandrel 5–10 times the diameter of the filament. The steel wire being subsequently dissolved by acid. In *coiled-coil lamps* the helical filament is wound

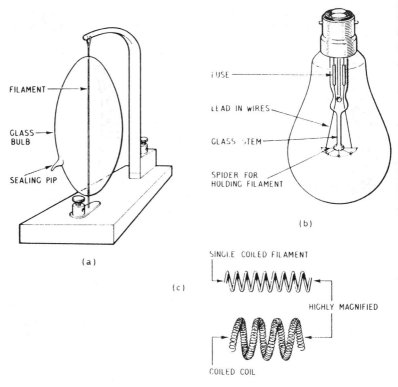

Fig. 15.13. (*a*) Early Swan carbon filament lamp (1878). (*b*) Tungsten lamp. (*c*) Single and coiled-coil filaments.

into a number of turns. This allows a higher working temperature and increases the efficiency, in terms of lumens/watt, by a further 10–20%. The filament fuse is a valuable feature as it avoids possible disruptive arcing and the blowing of circuit fuses when a lamp fails.

15.11. The Tungsten Halogen Lamp

Formerly termed the tungsten-iodine lamp, this type with filament enclosed in a small quartz bulb shows a 30% increase in efficiency and longer life over the conventional tungsten lamp. In spite of all improvements the tungsten lamp reveals a blackening of the lamp glass after a period of use. As already stated, this is due to the particles of the filament breaking off under the intense filament heat and being carried by convection currents to the lamp walls. In the tungsten-halogen lamp, the blackening action is

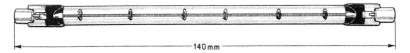

Fig. 15.14. 500 W tunsten halogen lamp.

prevented by adding a trace of iodine or bromine vapour to the filling gas. This sets up a regenerative cycle by which any evaporated tungsten is restored from the bulb wall *back to the hot filament.* The illumination is in the form of an intense white light. Linear lamps of this type (Fig. 15.14) with contacts at each end are available in 500–1500 W sizes, and so find a field of application in outdoor lighting. The relative small dimensions have brought the need for special design in caps, holders, springs, casing cover, glasses and gaskets to deal with the heat concentration.

15.12. Discharge Lighting—Basic Ideas

Discharge lamps of which there are two kinds, (*a*) hot cathode (*b*) cold cathode, developed from a series of highly scientific experiments designed to investigate the ultimate nature of matter. Towards the end of the last century, Sir J. J. Thomson was able to state the electron theory which forms the basis of electrical engineering science.

When an appropriate voltage is applied to two electrodes sealed into the ends of a tube containing gas, some interesting features appear. In a simplified explanation, the gas may at first be regarded as an insulator because the movement of any free electrons from the electrodes is hampered by a crowd of neutral atoms. If the pressure within the tube is reduced, the reduced gas pressure now allows increased electron movement which thus constitutes a current flow. As the atoms are now thinned out, the electron speed is sufficient to *excite* an atom, i.e. to knock other electrons from their orbits so that the atoms become ionised. This is accompanied by radiation producing the sensation of light if its wave length lies within the visible spectrum (Fig. 15.15). The electrons now collide with other gaseous atoms removing yet further electrons.

Increased electron velocity can be obtained by a higher voltage. The bombardment or avalanche brings light emission. As this chain reaction produces unstable conditions some form of resistance or choke must be placed in the circuit and connected in series with the tube.

On alternating current, the electrodes act alternately as cathode and anode every half cycle. To increase the flow of electrons, the

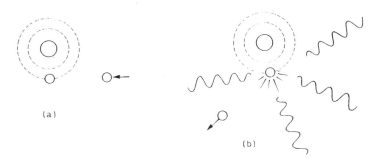

Fig. 15.15. (a) Electron at high speed approaching an orbiting electron of a gas atom. (b) A direct hit is made on the orbital electron. The atom is now excited and is accompanied by radiation.

electrodes are specially heated in all types of discharge lamp except cold-cathode types which are run at higher voltage as a substitute for heating. A coil of tungsten wire is used for cathode heating and the wire is chemically-treated to increase electron emission still further. All electrodes are heated by electron bombardment during discharge, but in some types of hot-cathode lamp the electrodes are preheated by passing current through them before the discharge starts. These tubes therefore need four lead-in wires instead of two.

15.13. The Hot-cathode Fluorescent Lamp

In this type of lamp the glass tube contains a drop of mercury which vapourises when heated, and a small amount of argon gas is introduced to assist starting. The action of passing a.c. through the tube produces invisible ultra-violet rays and a little blue light. The rays striking against the internally coated fluorescent powder change the frequency of the ultra-violet rays into radiations within the visible spectrum. The output light colour rendering depends upon the chemicals used for the fluorescent coating.

In the standard tube, the normal mains voltage would not be sufficient to initiate the discharge without an external start switch or starting transformer for pre-heating the cathode filaments. Figure 15.16 shows the glow-start switch method of starting. When the supply is switched on, the gases in the bulb of the start switch are heated by a glow-discharge across the open bimetal contacts so that the contacts 'make'. This allows the highly electron-emissive electrodes in the lamp to be pre-heated while at the same time an intense magnetic field is built up in the ballast choke. As the

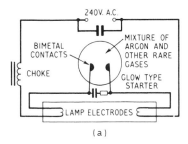

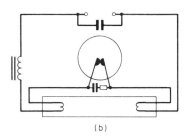

(a)   (b)

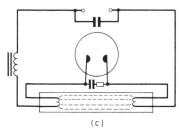

(c)

Fig. 15.16. Stages in the striking of a fluorescent tube, using glow-type starter switch. (a) Glow discharge between bimetal contacts; (b) Bimetal heated; contacts close, bimetal then cools. (c) Contacts snap open, causing arc to strike between electrodes.

contacts are closed, no current passes through the starter gases so that they cool and the starter contacts snap open, thus causing the choke field to collapse and thereby induce a momentary voltage of 500–1000 V for striking the arc across the electrodes. Should the tube fail to strike the cycle will repeat itself until the tube does strike. The starter, which has a small capacitor across the contacts for radio suppression, is now by-passed, while the choke serves its second function of acting as a stabilising ballast. The purpose of the capacitor across the mains is to improve the low power factor caused by the series choke. To comply with I.E.E. Regulations the p.f. capacitor requires a leak resistance to prevent any charges remaining on the plates.

The glow-type starter, as described above, has the advantage of requiring only two contacts. The contacts of the 4-pin thermal-type starter are initially closed, and only snap open when heated by a small heater element.

In contrast to the 1000-hour rated life of tungsten filament lamps, present-day fluorescent tubes claim an average life of 7500 hours, although a slight loss of electron-emissive material occurs whenever switching takes place.

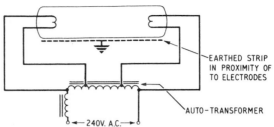

Fig. 15.17. Auto-transformer quick-start circuit.

15.14. Quick Starting

Much ingenuity has gone into preventing the irritating flicker at starting and eliminating the starter switch. One circuit is illustrated in Fig. 15.17. When the lamp is switched on, the end turns of the auto-transformer supply a current for heating the electrodes. As the main windings of the transformer are across the supply, quick starting is assured. The starting is assisted by the earthed strip. Since the neutral and earth are at the same potential, the gap between the live and neutral conductors at the ends of the tube are, in effect, much shortened so that the difficulties associated with the striking of the discharge are reduced. The earth strip may be in the form of a metal strip cemented along the length of the lamp.

15.15. Stroboscopic Effect

In the a.c. cycle zero value occurs twice per cycle and theoretically lamps should go out twice per cycle or every hundredth part of a second on a standard supply. In the tungsten lamp, the heating of the filament results in a 'carrying over' of this null period and therefore prevents lamps from being extinguished or even flickering.

Discharge lighting does not operate on the heated-filament principle, and although flickering may not be noticeable under normal circumstances, peculiar dangers—referred to as the stroboscopic effect—can arise where moving machinery is present. Spokes of a wheel, when rotating at the same speed as that of the supply, appear to stand still.

If the interval of time between the zero points is exactly equal to the time taken for one revolution of the wheel, the object will appear to be stationary because the *same spoke will be illuminated while in the same position each time.* The wheel will appear to be revolving backwards if the intervals of time are slightly shorter. If the intervals are longer the wheel will appear to be moving slowly forwards. This effect can be observed on the coach wheels of early films.

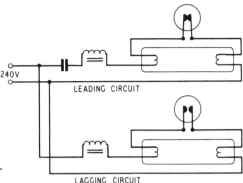

Fig. 15.18. The anti-strobo-
scopic lag-lead circuit.

240V

LEADING CIRCUIT

LAGGING CIRCUIT

There are now several solutions to this problem. Modern fluorescent powders possess an 'after-glow' effect to minimise the dark period. Twin fluorescent fittings can be arranged so that there is a 90° phase difference between the tubes (Fig. 15.18), the capacitor in one lamp circuit producing the phase lead. On large installations it is sound practice to connect fluorescent fittings on different phases where a 3-phase supply is available.

15.16. D.C. Operation

Fluorescent tubes will work on a d.c. but not so efficiently, the choke being replaced by a resistor, although it may be re-introduced as an additional item to give an initial impulse for ease in starting. After a time the tube darkens at the positive end due to the migration of ionised mercury vapour to the negative pole. To prevent this loss of light from increasing, a changeover switch must be fitted. The rotary switch has the advantage of ensuring polarity reversal every time the tube is switched on.

15.17. Cold-cathode Fluorescent Tubes

The general layout, when used for illumination purposes may be seen in Fig. 15.19. The 20 mm tubes are internally coated to give a wide variety of colours. The iron or nickel electrodes, which are cylindrically shaped, are not pre-heated as the tandem connected transformers supply continuous high voltage of 2–3000 V. The secondaries are centre-tapped to earth, while a capacitor is connected across each primary for power-factor improvement. The

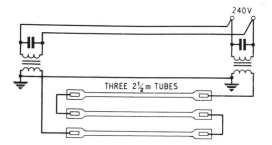

Fig. 15.19. Cold-cathode fluorescent circuit.

tubes, which have a long life of 15 000–30 000 hours, are often supplied in a variety of shapes for display purposes.

15.18. Sodium Lamp

The discharge tube (Fig. 15.20) is made of a special glass which is resistant to the sodium vapour, since ordinary quartz and hard glass are violently attacked by the sodium which vaporises at 300°C.

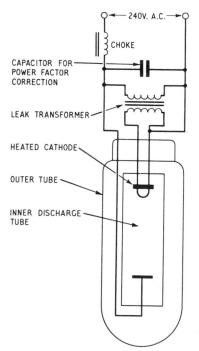

Fig. 15.20. Sodium lamp circuit.

The double glass container and high-reactance leak transformer are designed to assist in ease of starting as the discharge lamp would not start at the relatively low mains voltage. There is a vacuum between the two tubes for heat conservation. No starter switch is required as the transformer gives a starting voltage of about 400 V. The inner glass contains neon in addition to the metallic sodium at low pressure. Heat is produced by an initial neon discharge; at this stage a red light is emitted because of the neon gas. Due to the heated discharge the sodium then begins to vaporise causing the colour of the discharge to change from red to sodium yellow. It takes about 10 minutes for full light to be reached. The lamp is usually fitted with a b.c. holder of non-standard design.

A development has been the *high-pressure* sodium lamp. The efficiency reached (150 lm/W) is the highest yet attained by any electric lamp. Furthermore, at the high gas pressure used the spectrum of sodium is widened to give an adequate cover of all colours. This is in contrast to the yellow light of the older type which operates on a narrow waveband.

15.19. Mercury Vapour Lamp

Referring to Fig. 15.21(*a*), the discharge occurs in an inner tube enclosed in an outer evacuated tube. The inner tube, of glass or quartz, contains a little mercury and a small amount of argon to assist starting. Again in common with other hot-cathode discharge lamps, the electrodes are rich in electron-emitting materials in order to permit ease in the release of electrons.

The lamp operates at high pressure. To start a discharge an auxiliary electrode is positioned quite close to a main electrode, the auxiliary electrode being connected to the lamp terminal through a high resistor for limiting the current. On switching, the normal mains voltage is not sufficient to start the discharge between the main electrodes but it can start over the very short distance between the main and auxiliary electrodes. At this stage, the discharge is in the argon gas. The discharge current passing through the high resistance causes a p.d. to develop between the starting electrode and the main electrode. The discharge now spreads rapidly until it is taking place between the main electrodes. The argon discharge then warms up the tube and vaporises the mercury. Soon the gas content is mainly mercury vapour and the argon has less and less effect. The discharge then takes place in the mercury vapour.

The characteristic blue light renders the lamp impracticable for industry. An early attempt for colour correction was brought about

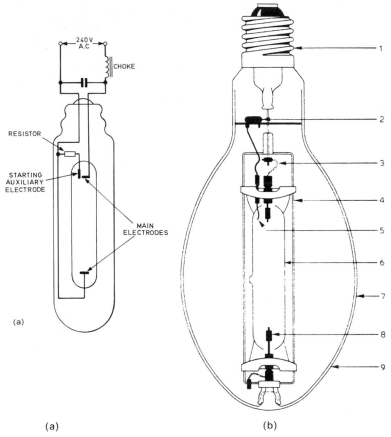

Fig. 15.21. Mercury vapour lamp. (*a*) basic circuit. (*b*) modern mercury vapour lamp.

1. Non-bending nickel-plated cap.
2. Starting electrode resistor. Limits current to a low value in the starting electrode during normal lamp running conditions.
3. Electrical connectors. Flexible braided connectors to eliminate failure by vibration.
4. Support assembly. Nickel plated to reduce light absorption.
5. Starting electrode.
6. Quartz arc tube. Degassed of all organic matter to give increased light and better starting characteristics.

7. Fluorescent coating. The special phosphor used gives a red ratio of 9% for optimum colour.
8. Main electrode. This electrode is of new design to hold a greater amount of emissive material, giving even a longer and more reliable life.
9. Outer envelope. Made of hard glass fully resistant to water and corrosive atmospheres.

by 'blended' lighting, being a dual arrangement by which an ordinary filament lamp was placed in the same bulb as the discharge lamp.

An improved modern type of mercury-vapour lamp, which gives a better colour rendering, is illustrated in Fig. 15.21(b). The improvement in colour is brought about by the internal fluorescent coating on the outer bulb. Whilst the efficiency does not reach that of the new sodium lamp it is still high, being as much at 50 lm/W. As is common with discharge lamps there is a long life. The ranges available are 80 W to 1000 W.

15.20. Discharge Lighting Regulations (Low voltage)

(1) SWITCH RATINGS. Due to the inductive load of the choke the switch rating must be double that of the steady load. Thus 5 A switches are only permitted to break a 2·5 A discharge-lighting load unless specially designed for inductive circuits.

(2) LOADING. The kVA demand may be estimated from rated lamp wattage × 1·8, provided the power factor is not less than 0·85 lagging. This allows for losses in the control gear.

(3) SCREENING. All live parts must be provided with proper screens of earthed metal and insulating material must be of adequate mechanical strength. Alternatively for installations on the exterior of buildings, the live parts are to be accessible only to authorised personnel. Glass screens are permitted, but again only where accessible to authorised persons. Insulated screens must be non-hygroscopic, anti-tracking and substantially incombustible.

15.21. Discharge Lighting Regulations (Above low voltage)

(1) MAXIMUM VOLTAGE–5000 V TO EARTH. Thus a 10 kV display unit could be used provided its supply transformer is centre-tapped to earth.

(2) H.V. TRANSFORMERS. Where the input exceeds 500 W, means are to be provided for automatic supply disconnexion in the event of a short-circuit or earth-leakage current which exceeds 20% of the normal steady current.

(3) INSTALLATION. All equipment to be housed in an earthed metal or substantial container suitable for high voltage. A notice 'DANGER—HIGH VOLTAGE' with type of lettering as stated in Regulation 554–5 to be permanently fixed near to the equipment.

For connexion to high-voltage cables, the insulation which is exposed by removing the metal sheath or braid must be protected from the effects of the sun's ultra-violet rays. Armouring may be required for protection against mechanical damage. Non-armoured cables may only be drawn through short *earthed* lengths of metal tubing where passing through walls or ceilings. Unless easily identifiable, high-voltage cables require red on white background 'DANGER' notices at intervals not greater than 1·5 m, the minimum height of the letters to be 8 mm.

(4) SEPARATION AND ISOLATION. High-voltage discharge lamps to be supplied via a double wound transformer. However auto-transformers may be used on 2-wire circuits which do not exceed 1·5 kV, one pole to be connected to earth and the control switch to be of the double pole type.

Isolation of live conductors may be made by either of the following methods.

(*i*) An interlock on self-contained fitting to be provided in addition to the switch normally used for controlling the circuit.

(*ii*) Local isolation by plug and socket or similar method in addition to normal control switch.

(*iii*) Switch with removable handle. Alternatively, a switch or distribution board of a type that can be locked may be fitted if the keys are held by authorised persons. Where there are more than one switch or distribution board all removable handles and keys to be non-interchangeable.

(5) FIREMAN'S SWITCH. These emergency switches, which are required for neon lighting installations, must be provided for all exterior h.v. installations including closed markets and arcades. Internally, their use is confined to circuits that are run unattended, such as in window-display signs. The Fireman's switch must also conform to the following requirements

(*i*) The switch must be able to isolate all live conductors. One switch only should be fitted to control the whole of an exterior installation and one for the interior.

(*ii*) It should be coloured red and an adjoining nameplate to be marked 'FIREMAN'S SWITCH' with regulation lettering.

(*iii*) The ON and OFF positions to be clearly indicated by lettering legible to a person standing on the ground; the switch *OFF position to be on the top*, and construction must be such as to prevent accidental movement to the ON position.

(*iv*) The Fireman's switch to be in a conspicuous position, which may be agreed with the local fire-brigade authority, at not more than 2·75 m from the ground. It is desirable that the switch be adjacent to the discharge lamps of an exterior installation and at the main entrance to the building for an interior installation.

EXERCISES

1. (*i*) State and explain the various factors to be considered when planning the electric lighting of *either*: (*a*) a large garage workshop, or (*b*) a workshop with a number of bench sewing machines.

(*ii*) an office 40 m by 20 m requires an average illumination at desk level of approximately 150 lux. The following alternatives are available, using (*a*) 80 W fluorescent lamps rated at 2800 lumens, (*b*) 100 W tungsten filament lamps rated at 100 cd.

Calculate the number of lamps needed in each case, assuming the coefficient of utilisation of the room to be 0·55, and the lamp depreciation factor to be 1·25.   [C as modified]

2. Two incandescent filament lamps with reflectors are suspended 2 m apart and 1·8 m above a level workbench. The reflectors are such that each lamp has luminous intensity of 200 candelas in all directions below the horizontal. Calculate the total illumination at bench level, immediately below each lamp and mid-way between them.

Describe briefly an instrument you could use to check the illumination in such a case. Explain how it works.   [C]

3. A small assembly workshop, 25 m by 18 m, requires a general illumination at bench level of 150 lux. The following alternatives are suggested: (*a*) 150 W tungsten-filament lamps giving 13 lumens per watt, (*b*) 80 W fluorescent lamps giving 35 lumens per watt.

(*i*) Assuming the lamp maintenance factor in each case to be 0·8 and the cofficient of utilisation of the room to be 0·6, calculate the number of lamps to be installed for both methods (*a*) and (*b*).

(*ii*) Briefly describe the types of lamp and lighting fitments you would fit in one of the following situations, giving reasons: (*c*) a large machine-shop with rows of lathes, drilling machines, etc., (*d*) a large drawing-office.
                                                                                      [C]

4. An incandescent filament lamp is suspended 1·8 m above a level work bench. The lamp is fitted with a reflector such that the luminous intensity in all directions below the horizontal is 400 candelas.

Calculate the illumination at a point A on the surface of the bench immediately below the lamp, and at other bench positions 1 m, 2 m and 3 m from A in a straight line. Show the values on a suitable diagram.

5. Describe clearly the general principles of operation of a mercury-vapour discharge lamp. Explain the differences between high-voltage and low-voltage lamps. Give reasons why a choke is necessary for the operation of a discharge lamp.

6. (*a*) Make a diagram showing the starting connexions for an ordinary fluorescent lamp connected to an a.c. supply. Explain the action of starting the lamp and describe the function of each separate part of the circuit in the starting operation.

(*b*) Show how an ordinary fluorescent lamp should be connected for use on a d.c. supply.

7. (a) A filament lamp and reflector having a distribution as tabulated below are suspended 2 m above the centre of a circular table of 4 m diameter. Using the data given below, calculate the illuminance on the table: (i) directly below the lamp (ii) at the circumference of the table.

Angle in degrees measured through centre of lamp cap	0	15	30	45	60	75
Luminous intensity (cd)	280	310	400	460	475	420

(b) Explain why large filament lamps are gas filled.

8. (a) A luminaire producing a luminous intensity of 1600 cd in all directions below the horizontal is suspended 8 m above a level bench. Calculate the illuminance on the surface:

(i) directly below the lamp.

(ii) at a point 4 m along the bench from (i).

(b) A workshop 30 m long and 15 m wide is to be provided with an illuminance at bench level of 300 lux using 125 W fluorescent lamps of 35 lumen/W efficacy. Assuming a utilisation factor of 0·5 and a maintenance factor of 0·8 calculate:

(i) the power required.

(ii) the number of lamps required.

Chapter 16

HEATING

16.1. General Principles of Temperature and Heat

The heat content of a material or body depends upon
(a) temperature,
(b) weight of material, and
(c) type of material

Thus, temperature and heat are not the same; nor can heat content be measured by a thermometer. Temperature is indicative of the *level* of heat—a measure on an agreed scale of the ability of heat to transfer from one substance to another, or from one part of a substance to another part of the same substance.

The two main temperature scales, Fahrenheit and Celsius are connected by

$$°F = (\tfrac{9}{5} \times °C) + 32$$

$$°C = (°F - 32)\tfrac{5}{9}$$

These are awkward formulae and easy to forget. The author prefers an infallible method which does not rely on blind memory. The boiling point of water on the Fahrenheit (212°) and Celsius (100°) scales are known. Clearly, subtracting 32 from 212 and then multiplying by 5/9 gives 100. Similarly, adding 32 to the product of 100 and 9/5 brings 212. Freezing points on the Celsius and Fahrenheit scales are 0° and 32° respectively, below zero values the temperature has a negative sign.

Example 16.1

Convert
(a) (i) 212°C to °F, (ii) −12°C to °F;
(b) (i) 100°F to °C, (ii) −34°F to °C;
(c) *A Fahrenheit thermometer indicates a rise of 20°F when placed in a liquid. What rise would be shown on the Celsius scale?*

(a)	(i)	$(212 \times \tfrac{9}{5}) + 32$	$= 381 \cdot 6 + 32$	$= 413 \cdot 6°F$
	(ii)	$(-12 \times \tfrac{9}{5}) + 32$	$= -21 \cdot 6 + 32$	$= \underline{10 \cdot 4°F}$
(b)	(i)	$(100 - 32)\tfrac{5}{9}$	$= 68 \times \tfrac{5}{9}$	$= 37 \cdot 8°C$
	(ii)	$(-34 - 32)\tfrac{5}{9}$	$= -66 \times \tfrac{5}{9}$	$= \underline{-36 \cdot 7°C}$

(c) Since one degree F is equivalent to $\frac{5}{9}$ degree C, the given rise is equivalent to $20 \times \frac{5}{9}$ or $9\frac{1}{9}$ degrees C.

16.2. Heat Units

Calorie. The heat required to raise 1 gram (g) of water 1°C. In SI units, the JOULE is taken as the unit of heat.

$$
\begin{aligned}
1 \text{ calorie} &= 4\cdot19 \text{ joules} \\
1 \text{ W} &= 1 \text{ joule/second (J/s)} \\
1 \text{ Ws} &= 1 \text{ J} \\
1 \text{ kWh} &= 1000 \times 3600 \text{ J} \\
&= 3\cdot6 \times 10^6 \text{ J} \\
&= 3\cdot6 \text{ megajoules} \\
&= 3\cdot6 \text{ MJ}
\end{aligned}
$$

Volume or capacity of 1000 cm³ is equal to 1 litre

Mass of 1 litre of water = 1 kg

16.3. Specific Heat

Water requires 4190 joules to raise a mass of 1 kg of water 1°C. Other materials require different amounts of heat to raise 1 kg of their mass 1°C. Thus, 1 kg of brass requires only 390 joules to raise its temperature 1°C. Hence 390 J/kg°C is known as the specific heat (c) of brass.

The heat required to raise the temperature of a substance in joules

= mass (kg) × temp. change (°C) × sp.ht. (J/kg °C)

Example 16.2

A 3-kW immersion heater is fitted to a tank containing 113·65 litres of water. If the initial temperature of the water was 27°C, how long would it take to reach 77°C.
 (a) assuming no heat lost,
 (b) assuming an efficiency of 80 %.

(a) Electrical input = $3 \times t$ kWh where t is in hours. Since 1 kWh = $3\cdot6 \times 10^6$ J and 1 litre of water weighs 1 kg.

Heat output = *wt (kg) × temp. change (°C) × specific heat water*

$$(J/kg°C)$$

$$= \frac{113\cdot65 \times 50 \times 4190}{3\cdot6 \times 10^6} = 6\cdot6 \text{ kWh}$$

The electrical input and heat output can be equated as they are now both expressed in the same units.

$$3t = 6\cdot6$$

$$t = \frac{6\cdot6}{3} = 2\cdot2 \qquad = \text{2 hours 12 minutes}$$

(b) Efficiency $= \dfrac{\text{output}}{\text{input}}$.

$$\frac{80}{100} = \frac{6\cdot6}{3t}$$

$$t = \frac{6\cdot6 \times 100}{3 \times 80} = 2\tfrac{3}{4} = \text{2 hours 45 minutes}$$

Example 16.3

A number of steel plates, of mass 5 kg in all, are to be heated in a 240 V a.c. furnace from a temperature of 33·3°C to the annealing temperature of 750°C. Calculate the efficiency of the furnace, if the average current taken is 3·9 A at unity power factor, and the operation time is 40 minutes.

The specific heat of steel may be taken as 511 J/kg°C

$$\text{Input} = \frac{VI \times t\,(\text{hours})}{1000} = \frac{240 \times 3\cdot9 \times 40}{60 \times 1000} \qquad = 0\cdot624\,\text{kWh}$$

$$\text{Output} = \text{mass} \times \text{temp. change} \times c = 5 \times (750 - 33\cdot3) \times 511\,\text{J}$$

$$= \frac{5 \times 716\cdot7 \times 511}{3\cdot6 \times 10^{6}}\,\text{kWh} \qquad = 0\cdot5087\,\text{kWh}$$

$$\text{Efficiency} = \frac{\text{output}}{\text{input}} = \frac{0\cdot5087}{0\cdot624} \times 100 \qquad = 81\cdot5\%$$

Example 16.4

In a 240 V circuit a short-circuit occurs, the resistance being 1·5 Ω. If the circuit is wired in 1·5 mm² cable, calculate the tempera-ture (°C) rise after 5 seconds. The length of cable is 36·57 m and the mass/1000 m of 1·5 mm² copper is 13·21 kg. Specific heat of copper is 380 J/kg°C.

$$\text{Mass of copper} = \frac{13\cdot21}{1000} \times 36\cdot57 \qquad = 0\cdot4831\,\text{kg}$$

Heat energy in joules $=$ mass (kg) $\times T \times c$ where $T =$ temp. rise in $^\circ$C

$c =$ specific heat

$$= 0{\cdot}4831 \times T \times 380 = 183{\cdot}5T \qquad (1)$$

$$\text{Fault current} = \frac{V}{R} = \frac{240}{1{\cdot}5} \qquad\qquad = 160\,\text{A}$$

Electrical energy in joules $= I^2Rt$ where $t =$ time in seconds

$$= 160^2 \times 1{\cdot}5 \times 5 \qquad\qquad = 192\,000$$

$$(2)$$

Equating (1) and (2)

$$183{\cdot}5T = 192\,000$$

$$\text{Temperature rise} \qquad T = \frac{192\,000}{183{\cdot}5} \qquad\qquad = \underline{1046^\circ\text{C}}$$

which is just below the melting point (1084°C) of copper.

Example 16.5

A 600 kVA 3-phase transformer is immersed in a tank containing 10 000 litres of insulating oil. The efficiency of the transformer at full load is 97%.

Calculate the average rise in temperature in $^\circ$C of the oil after a 3-hour run at full load and unity power factor, assuming that 60% of the heat energy lost in the transformer is expended in heating the oil (specific heat of oil $= 2142\,J/kg^\circ C$, 1 litre of oil has a mass of 0·9 kg). [C as modified]

Output $= 600\,\text{kVA} = 600\,\text{kW}$ at unity power factor

$$\text{Efficiency} = \frac{\text{output}}{\text{input}}$$

$$\therefore \quad \frac{97}{100} = \frac{600}{\text{input}}$$

$$\text{input} = \frac{600 \times 100}{97}\,\text{kW}$$

But

$$\text{Losses} = \text{input} - \text{output} = \frac{600 \times 100}{97} - 600$$

$$= 600\left(\frac{100}{97} - 1\right) = 19\,\text{kW}$$

So that

Energy losses for 3-hour run $= 3 \times 19 = 57\,\text{kWh}$

As heat given to oil is 60% of heat energy lost in transformer

$$\text{Heat given to oil} = \frac{60 \times 57}{100} = 34 \cdot 2\,\text{kWh} \qquad (1)$$

Also
$$= \text{mass} \times \text{specific heat} \times \text{temp.}$$
$$\text{rise } (t) \text{ joules}$$
$$= \frac{10\,000 \times 0 \cdot 9 \times 2124 \times t}{3 \cdot 6 \times 10^6}\,\text{kWh}$$
$$= 5 \cdot 31t \qquad (2)$$

Equating (1) and (2) $34 \cdot 2 = 5 \cdot 31t$

$$t = \frac{34 \cdot 2}{5 \cdot 31} \qquad\qquad = \underline{6 \cdot 4^\circ\text{C}}$$

16.4. Calorific Value of Fuels

Apart from electrical methods, heat is usually produced by the combustion or burning of fuels. Thus fuels such as wood, coal, coal-gas and oil contain chemical energy, which when combining with the oxygen of the air give out heat energy. The number of heat units/kg (J/kg) of a fuel is termed its calorific value.

Example 16.6

A bath requiring 100 litres of water has to be heated from 25°C to 61°C.

(a) If the calorific value of coal is $33 \cdot 5 \times 10^6$ J/kg, how much coal would be required?

(b) Calculate the electrical energy which could be produced by this same amount of coal.

Neglect losses. Take the specific heat of water as equal to 4200 J/kg°C.

(a) Heat required = mass of water × temperature change × specific heat. Since the mass of 1 litre of water is 1 kg

$$\text{Heat required} = 100 \times (61 - 25) \times 4200 \qquad = 15 \cdot 12 \times 10^6\,\text{J}$$

$$\text{Coal required} = \frac{15 \cdot 12 \times 10^6}{33 \cdot 5 \times 10^6} \qquad = \underline{0 \cdot 45\,\text{kg}}$$

(b) $1 \text{ kWh} = 3.6 \times 10^6 \text{ J}$

$\therefore$ Equivalent electrical energy at 100% efficiency

$$= \frac{15.12 \times 10^6}{3.6 \times 10^6} \qquad = 4.2 \text{ kWh}$$

16.5. Heat Transference

Heat may be said to flow from a hot body towards a cold body. A heated substance gives off heat to another substance by (a) conduction, (b) convection, (c) radiation. The various types of electrical heaters operate on one or more of these principles.

CONDUCTION. In conduction there is direct heat transference through a substance from one point to another. In a heated rod, for example, heat energy is transmitted from molecule to molecule by direct contact, although there is no movement of the molecules themselves. Atoms in any material are in constant vibration, this vibration being increased by any rise in temperature; most substances appear to follow this pattern. Copper is a good heat and electrical conductor, while paper is a good heat and electrical insulator.

CONVECTORS. The use of convection currents is probably the most important means of transmitting heat energy for obtaining both space and water heating. Air itself it not a good conductor but the layer of air in contact with a heated element is given heat energy and thereby expands. Due to expansion the density of this

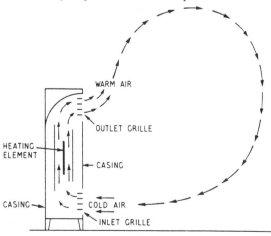

Fig. 16.1. Principle of the electric convector.

air is reduced so that it becomes lighter and rises. A fresh layer of air takes its place and in turn rises. By this means continuous circulation of heated air may be obtained, the principle of which is illustrated in Fig. 16.1. Cool air is drawn into the electric convector at ground level, warmed by the heat and discharged through the top grille. By operating at black heat the life of the nickel-chrome element is increased as compared with radiators where they run red hot. Convection heating, which may be fan-assisted, lends itself to thermostatic control but, when fitted, the thermostat should be so placed as to respond to the temperature of the incoming air flow.

Tubular heaters also act as convectors. They are sheet-steel cased of circular (50 mm diameter) or oval section containing an element and may be obtained in lengths from 0·61 m to 5·2 m (Fig. 16.2(*a*)). Brackets maintain a wall clearance of 33 mm. Interconnectors between the heaters are obtainable and to provide a compact loading the heaters are often fitted in tiers. The flexibility offered by the range of sizes makes them very effective for checking draughts. Cold downdraughts of air from skylights and windows may be warmed by fixing appropriate tubular heaters below the glass line.

As a more modern approach the units may be modified or incorporated in skirting heaters which are finished with attractive colours to blend in with present-day home or office decor. One form is shown in Fig. 16.2(*b*).

RADIATION. The sun warms the earth by radiation which travels with the speed of light. In the electric radiator (Fig. 16.3), heat rays from an element at bright red heat pass rapidly through the air without heating the atmosphere but raise the temperature of solid substances, within its range. Thus walls, furniture and human bodies absorb the rays and become hotter. As can be seen by the sketch the heat rays travel in straight lines and are reflected by polished surfaces, in the same way as light, but are absorbed by matt black finishes. Reflected radiator heaters are *not* suitable for thermostatic control.

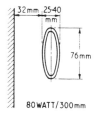

80 WATT/300mm

(a)

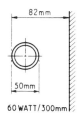

60 WATT/300mm

Fig. 16.2. (*a*) Tubular heater dimensions.

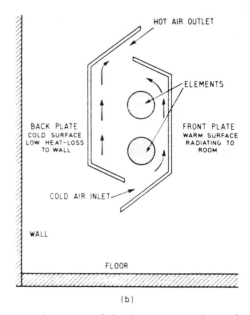

Fig. 16.2. (*b*) Tubular heaters incorporated in skirting heaters.

(b)

It is also interesting to note that part of the heat passes through the reflector by conduction which in turn heats layers of surrounding air to produce a certain amount of convected heat.

To resist oxidisation, elements are wound with a nickel-chrome (80% nickel 20% chrome) wire with a small temperature co-efficient. This alloy has the additional advantage of high resistance per unit length so that a compact heat source is obtained.

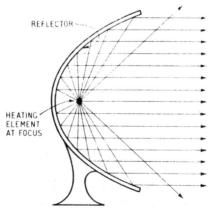

Fig. 16.3. Parabolic radiator reflector.

Infra-red ray heaters may be designed to send out long electro-magnetic waves in the 3 micron range to produce bodily comfort. The heater element is enclosed in a fused silica tube. This material is a bad conductor but highly transparent to the infra-red radiations. Furthermore, the tube acts as a protection from irregular air current striking the elements and shortening their life. The refractory tube material also provides a degree of safety in that it prevents the heated wire from being touched.

An almost endless variety of electric fires functions on one or more of the basic principles of heat transference. Mention may be made of *industrial unit heaters* which range up to several kW. Wound with a coiled wire element which is placed in front of a fan, they may be mounted at a height of some 2–3 m and blow hot air over a wide area. In this way a degree of forced ventilation is also provided. The *oil-filled radiator* is another heater with both radiant and convective output. Originally designed in the shape of the central-heating column-type radiator, they are now often dimpled and possess a neat pressed-steel slimline casing. The unit is partially filled with a high-grade oil which is heated by an immersion heater; the oil expands and evenly heats the casing. There is a built-in thermostatic control, with an overload cut-out for the larger sizes, and the heater may be wall-mounted or free-standing.

Panel heaters are also produced in a variety of types. Flat elements are embedded in various materials. In one type, a carbon element is sandwiched between slate material panels. Most of the heat is given off as low-temperature radiation. It has been found that a large proportion of radiant heat is delivered when the panels are fitted to the ceiling.

16.6. Room Heating Calculations

Many factors have to be taken into account and involve a know-ledge of:

(1) Number of air changes required per hour. The replacement of stale air with fresh requires additional heat.
(2) Area of windows, doors and walls.
(3) Situation of walls, i.e. whether internal or external.
(4) Exposure of ceiling. Allowance may have to be made accord-ing to whether there is a roof or heated room above.
(5) Materials from which the walls, floor and ceiling are made.
(6) Internal temperature required.
(7) Outside air temperature.

Example 16.7

Low-temperature heating panels with a total load of 8 kW at 240 V are used to heat a room 10 m × 6 m × 4 m. Of the heat generated 40% will be used to heat the air in the room. Calculate the average temperature rise if there are two changes/hour. Density of air = 1·2 kg/m³. Specific heat of air = 1010 J/kg°C.   [C & G]

$$\text{Volume of air} = 10\,m \times 6\,m \times 4\,m \qquad = 240\,m^3$$

$$\text{Mass of air} = \text{volume} \times \text{density} = 240 \times 1\cdot2 = 288\,kg$$

Mass of air changed/hour

$$= 288 \times 2 \qquad = 576\,kg$$

Heat energy required $=$ mass $\times$ temp. rise (t) $\times$ specific heat

$$= 576 \times t \times 1010\,J$$

$$= \frac{576 \times 1010t}{3\cdot6 \times 10^6} \qquad = 0\cdot1616t\,kWh \quad (1)$$

Useful electrical energy

$$= 8 \times \frac{40}{100} \qquad = 3\cdot2\,kWh \quad (2)$$

Equating (1) and (2)

$$0\cdot1616t = 3\cdot2$$

$$t = \frac{3\cdot2}{0\cdot1616} \qquad = 19\cdot8°C$$

∴ $\underline{\text{Temperature rise} = 19\cdot8°C}$

16.7. Transmission Coefficients (U-values)

When designing any space heating installation, apparatus must be installed which will make up for heat losses from the body. This follows from fundamentals that the direction of heat flow is from hot to cold. For this purpose, heating engineers make use of heat transmission coefficients or U-values. They represent the quantity of heat in joules that will pass in one second through $1 m^2$ of any item of building construction when the temperature difference between the two sides of the material is $1°C$. As 1 joule/second = 1 watt:

Heat loss in watts = area $(m^2) \times$ U-value $(W/m^2 °C) \times$ temperature difference $(°C)$

A comprehensive list of transmission coefficients has been prepared by the Institution of Heating and Ventilating Engineers.

Example 16.8

A small workshop, $13 \cdot 5 m$ by $6 \cdot 5 m$ with a ceiling height of $4 m$, has a window area of $36 m^2$. The workshop is to be heated electrically so that the average temperature inside is maintained at $21°C$ when the temperature outside is $0°C$.

Calculate the required power in kW on the assumption that there will be two complete changes of air per hour.

Density of air = $1 \cdot 2 kg/m^3$. Specific heat of air = $1010 J/kg°C$
Heat transmission coefficients, $W/m^2 °C$:

Walls, brick (cavity construction), inside plastered	1·76
floor, concrete	11·5
ceiling, concrete, asphalt-covered, expanded polystyrene underside, inside plastered	0·971
window, glass	5·69

SOLUTION

Areas		m^2
Gross area of walls	$= 2[(13 \cdot 5 \times 4) + (6 \cdot 5 \times 4)]$	$= 160$
Window		$= 36$
Wall area to be heated	$= 160 - 36$	$= 124$
Area of ceiling	$= 13 \cdot 5 \times 6 \cdot 5$	$= 87 \cdot 75$
Area of floor	$= 13 \cdot 5 \times 6 \cdot 5$	$= 87 \cdot 75$
Heat Loss	$=$ Area $\times$ U-value $\times$ temp. diff.	*watts*
Walls	$124 \times 1 \cdot 76 \times 21$	$= 4\ 583$
Window	$36 \times 5 \cdot 69 \times 21$	$= 4\ 302$
Ceiling	$87 \cdot 75 \times 0 \cdot 971 \times 21$	$= 1\ 789$
Floor	$87 \cdot 75 \times 11 \cdot 5 \times 21$	$= 21\ 192$
		$31\ 866$

Volume of air required for two air changes

$$= 2 \times 13\cdot5 \times 6\cdot5 \times 4 \qquad = 702 \text{ m}^3$$

$$\text{Mass of air} = 702 \times 1\cdot2 \qquad = 842\cdot4 \text{ kg}$$

Mass of air changed/second

$$= 842\cdot4 \div 3600 \qquad = 0\cdot234 \text{ kg/s}$$

Heat loss = mass of air changed/s × temp. diff. × specific heat

$$= 0\cdot234 \times 21 \times 1010 \qquad = 4\,963 \text{ W}$$

Heat loss as obtained above $\qquad = 31\,866 \text{ W}$

Total watts $= 36\,829$

Power required $= 36\cdot8$ kW

16.8. Off-peak Heating Installations

Off-peak heating is invariably based on the heat-storage principle. During the night the heater stores up heat until either its thermostat operates or a time-switch switches off the supply. A cheap rate applies and there may be a mid-day boost. The four main methods employed are: (i) unit block storage heaters, (ii) floor warming, (iii) air-duct heating and (iv) the use of water as the storage medium.

In the uncontrolled type of block storage heater, a wire-wound element is embedded in fireclay, concrete, or a similar refractory material having a high thermal capacity for storing heat, and enclosed in a metal casing. The casing may be lined with a heat-resisting material such as glass fibre. There may be a built-in thermostat to allow the consumer to control the charge input to the heater.

The controlled type of block storage heater is fan-assisted (Fig. 16·4), the fan being switched on when a rapid discharge of warm air is required to circulate round the room. The fan is connected to the normal unrestricted supply. Additional output control can be obtained by means of an adjustable damper. A room thermostat may be fitted and connected in the fan circuit to provide an automatic on–off cycle of operation throughout the day; as the temperature of the storage block falls, the fan runs for longer periods. The room thermostat may be subject to the over-riding control of a time switch for bringing the fan into action under thermostatic control at predetermined times during the day. The heater contains an adjustable input charge thermostat and, to prevent overheating, a thermal cut-out.

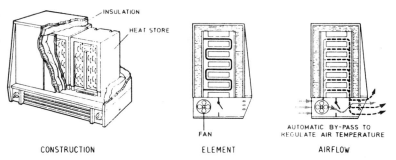

CONSTRUCTION   ELEMENT   AIRFLOW

Fig. 16.4. Fan controlled block storage heater.

Where the input charge to a storage-heating installation is to be automatically controlled in accordance with changing weather conditions, some form of external regulator, which is sensitive to outside temperature, is placed in series with the time switch for controlling the time of off-peak switch-on.

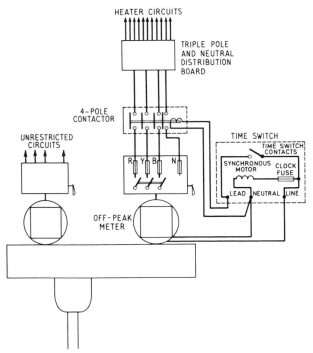

Fig. 16.5. 3-phase off-peak heating controls (meter tail connexions have been omitted for clarity).

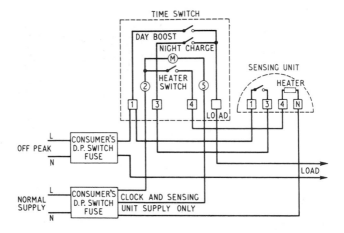

Fig. 16.6. Off-peak heater control with sensing unit.

The newer white-meter tariff is operated by a meter having two dials and a time switch, but off-peak wiring that is separately metered must constitute a completely separate installation. The general supply and consumer's control arrangement where heaters are spread over three phases is given in Fig. 16.5. There may be an additional consumer's time switch which is adjusted by the consumer so as to operate only for part of the off-peak period. Figure 16.6 illustrates an alternative scheme which incorporates a Sangamo–Weston external sensing unit. Consumer units have been developed which comprise unrestricted and off-peak control gear. Since the heaters are time-switch controlled, switch-sockets must not be used for the final flexible cord connexion. Should an exposed element radiator be plugged-in to the switch-socket when the time-switch has cut off the supply, the circuit may be mistaken as being disconnected. Under these circumstances, potential dangers arise when the time-switch cuts in. The normal plug and socket must be replaced by a fixed connexion preferably through a flex-outlet.

The linked double-pole twin switch (Fig. 16·7) meets the requirements for fan-controlled output storage heaters. A box should be fitted in accordance with I.E.E. Regulations as the barrier gives protection where the fan and heater are on different phases.

When calculating storage heater loads, no allowance should be made for diversity. Radial circuits are preferable but the fan circuit must be on the normal (non off-peak) supply.

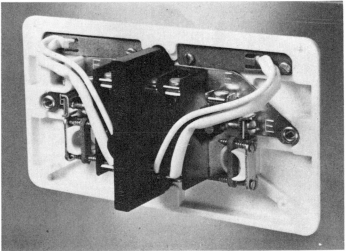

Fig. 16.7. Front and rear views of switch for fan controlled heater.

16.9 Floor Heating

This type of heating is applicable to solid concrete floors where
the mass of concrete possesses a high thermal capacity. Insulated
heating cables, of nylon, glass fibre insulation or other types are laid
in the structure of the floor to produce low-temperature radiation
heating. The cable is laid out in a continuous grid without any of the

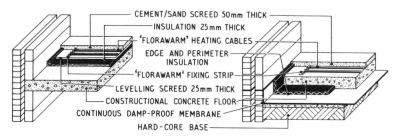

Fig. 16.8. Embedded floor heating cables.

wires crossing, loading of 15 watts per 0·1 m² being usually found adequate. To avoid discomfort, the mean floor temperature must not exceed 25–27°C. Thus, in some cases floor heating may be regarded as providing background heat so that some 'topping up' may be required by unit block storage or other types of heaters.

Since the cables are laid in the structure of the floor, close collaboration with the building contractors is essential. To conserve the maximum heat and avoid losses, insulation under the cables and along the room edges must be provided. The actual wiring is carried out by two main methods, *withdrawable* and *embedded*.

The first method is more costly and required ducts to be formed in the floor through which the cables are threaded. Embedded cables require a temporary jig to ensure that the grid of cables are maintained in position during laying and screeding, the latter being 50–75 mm thick. While the screed is being formed, monitoring of the cables to check for any breaks is essential to avoid the possibility of cutting out damage after the cement has set. The general scheme using Falk's 'Florawarm' cables may be seen from Fig. 16.8. The 'heater' may be regarded as a fixed appliance, although Regulations permit the switch of any particular section of cables to be in another room. The main controls with ancillary time-switch and contactor follow those as already given under unit storage heaters (Figs. 16.5–16.6).

16.10. Air Duct Heating

This is a form of electric central heating which is now commonly known as Electricaire and is a development of controlled-output block storage heating for domestic dwellings. Here the unit (Fig. 16.9) is centrally positioned in the building. Its storage core is made from refractory blocks, or in some cases cast iron, air channels being formed in the core. A fan drives warm air through the ducts which

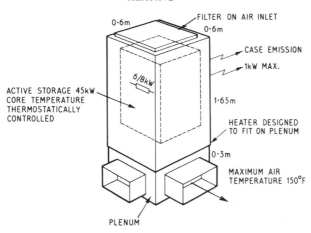

Fig. 16.9. 'Electricaire' air duct heater.

are distributed about the house with adjustable grille outlets at skirting level. As a refinement, there may be a 2-speed fan to bring a cold area rapidly up to the required temperature and an automatic warm-air mixing device.

WATER HEATING

Electricity can be made to provide cheap and plentiful supplies of hot water especially if obtained at off-peak rates. As most electric water heating is produced by immersion heaters, this type will be considered first.

16.11. The Immersion Heater

The tubular or withdrawable type (Fig. 16.10(a)) consists of resistance coils carried in slotted cylinders of ceramic material. It is sometimes known as the core type. The advantage of this type is that should a failure occur, the heating element can be renewed without disturbing the tank containing the water and without the necessity of draining down and recharging the vessel.

The element of the mineral-insulated, metal-sheathed non-withdrawable type (Fig. 16.10(b)) appears to be similar to the m.i.c.s. cable, as used for general wiring. To obtain a high electrical resistance for the comparatively short length, the element is made of

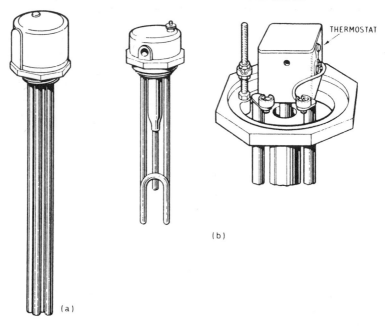

Fig. 16.10. Immersion heaters. (*a*) Withdrawable type. (*b*) The loop-type non-withdrawable immersion heater.

80/20 nickel-chrome resistance wire and is insulated by magnesium oxide, the whole being contained in a heavily plated seamless tube of deoxidised copper for maximum heat transference.

Siting. Since 'hot water rises', immersion heaters are often fixed low down in the tank. If fitted half-way, only half of the tank would be heated, the liquid below the heater being considered as 'dead water'. There should however be an allowance for a clearance of approximately 50 mm at the bottom of the tank, otherwise any scale or sludge formed if in contact with the immersion heater will certainly damage it.

Vertical fixing is normally from the top downwards, and has the advantage that small amounts of hot water, say for washing up, can be drawn off reasonably rapidly, this avoids the necessity for heating the whole of the tank as required when the heater is placed in the horizontal position. Some authorities criticise vertical fixing as hot water builds up from each section of the element, so that over-heating could occur at the top of the heater.

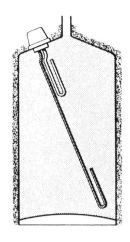

Fig. 16.11. Dual immersion heater.

Another possibility is the use of two immersion heaters fixed horizontally, one at low level and the other half-way up. The scheme is effective, if costly, for raising varying and rapid quantities of water. It has led to the development of the dual immersion heater (Fig. 16.11). In one type the short element, thermostatically controlled, heats approximately 26 to 45 litres and keeps it ready for domestic demand. Should a greater quantity of water be required, the long element, also thermostatically controlled, is switched on and sufficient water is usually available for a bath within an hour. By heating the water only as required, heat losses can be kept down to a minimum.

CIRCULATORS. These are types of immersion heater fitted in a similar position to that shown in Fig. 16.11 and, with the object of drawing off hot water rapidly, the element is shrouded by means of a vertical tube filled with holes. Difficulties can arise with hard water which is often responsible for completely choking the tube with scale.

16.12. Protection Against Scale

Hard water produces scale and sludge and affects the efficiency of the heating system. Contrary to general belief water, as found in nature, is seldom pure H_2O. Even water from clouds in the form of rain acquires salts and impurities from the air as it falls.

In streams and ponds, water soon becomes contaminated and dirty, and unfit to drink. By the time this water has filtered through to deep springs and wells it has been freed from suspended matter and organic contamination, but it has taken up considerable quantities of dissolved salts. These salts, consisting mainly of calcium and magnesium, account for the hardness of normal domestic water. When water is heated at the immersion-heater surface, carbon dioxide is dispelled and calcium carbonate is deposited. It is this deposit which is known as 'scale' or 'fur' and sometimes as boiler crust. The scale produced by hot surfaces, especially if excessive, periodically breaks, taking with it minute portions and layers of copper, hence the pitted appearance of elements that have become faulty in this manner.

Monel is an alloy consisting of nickel, copper, iron and manganese which, when used as the metal sheath for these heaters, withstands these conditions much more successfully.

Softening can be achieved with water of high-level temporary hardness by adding sufficient milk of lime. The salt then solidifies in the form of a chemical substance known as a precipitate and does no further harm.

16.13. Non-pressure Water Heater

The non-pressure water heater normally feeds one position at a time, i.e. bath, basin or sink, although with a swivel spout it can be used to supply two adjacent basins. The action is simple, as may be seen in Fig. 16.12. Hot water is obtained when cold water enters the water heater by turning on the tap. The inlet baffle is designed to prevent incoming cold water from mixing with the hot. The anti-drip syphon is necessary because water increases in bulk by some 0·2% per 10°C rise in temperature, therefore there would be constant dripping unless this or a similar device is fitted.

The usual 7–13·5 litre sizes are not only suitable for domestic premises but can also be used with advantage in small business establishments, where the requirements of the Offices and Shop Act require running hot water for the use of the staff. Fixings have to take account of the mass of the water, which is 1 kg for each litre. Installation is comparatively simple, it is only necessary to fix a short length of pipe from behind the existing cold water tap to the inlet valve of the heater. In some instances, the sink cold tap is screwed off and replaced by a tee-piece into which the tap is screwed off and replaced by a tee-piece into which the tap is fitted back again, leaving a spare opening for the heater. The connexion is sometimes carried out by means of small copper tubing.

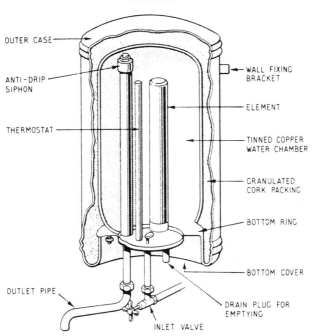

OUTER CASE

ANTI-DRIP
SIPHON

THERMOSTAT

OUTLET PIPE

INLET VALVE

WALL FIXING
BRACKET

ELEMENT

TINNED COPPER
WATER CHAMBER

GRANULATED
CORK PACKING

BOTTOM RING

BOTTOM COVER

DRAIN PLUG FOR
EMPTYING

Fig. 16.12. Single-point water heater.

Some makers supply a 'do-it-yourself' kit, the link being made by a piece of plastic pipe.

When used at the kitchen sink to avoid switching on a main immersion heater, a considerable economy can be effected. In addition, although the unit is heated by a small immersion heater, it is much more efficient than the tank-immersion-heater combination due to losses in long pipe runs.

16.14. Pressure-type Water Heater

This type is necessary when all the hot-water taps are supplied from one water heater. The word pressure here means that the outlet pipe may be closed and *not* that the heater can be supplied at mains pressure. In fact, it must be supplied from a cistern at an appropriate height, which must not exceed 20 m. The heater (Fig. 16.13) is fed from a ball valve cistern, usually situated in the loft in a similar way to the fuel-fired boiler. Note the hot service pipe from the top of the heater runs horizontally for a short distance before

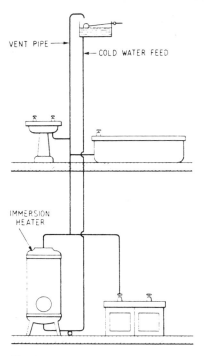

Fig. 16.13. Pressure-type water heater.

connexion to the vent pipe. This checks the 'one pipe' incipient circulation in the rising pipe during long idle periods, such as at night, which can cause a mixing of hot and cold water.

16.15. Off-peak Water Heating

Manufacturers have settled on the characteristic tall and slim vertical shape for the 225 litre tank (Fig. 16.14). The reason is that in the pressure type there is the tendency for the heated water to store in the upper part of the vessel—'hot water rises'—and thus to lose some of its heat to the cold water below it. This cold water as it flows to replace the hot water drawn off should not stir up and mix with the water above. In the ordinary hot water tank there is a considerable conduction of heat downwards by the metal sides of the vessel and there is also a small amount of transfer of heat between the strata or layers of hot and cold water. There are thus two losses

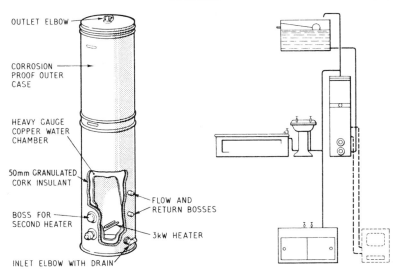

Fig. 16.14. 225-litre off-peak water heater. This type of heater must not be directly connected to the main water supply but must be fed from a ball-valve system.

of heat which can be reduced by a smaller circumference and therefore taller tank. The cold-water inlet is also well baffled to minimise the effects of hot and cold water mixing. Obviously, there must also be efficient lagging to prevent the temperature of the stored water from falling at a greater average rate than 0·55°C per hour.

16.16. The Electrode Boiler

To understand the principle of operation let us see what happens when two wires are placed close to each other in a container with water. Starting with direct current at a value as low as 3 volts, bubbles will be seen to collect at each wire. The bubbles contain gas and demonstrate that the water has been split up or decomposed into its constituent parts. Since chemically water is H_2O, oxygen gas bubbles collect at the positive wire or pole and twice as much hydrogen gas (H_2) evolves at the negative pole.

A thermometer placed in the water will show a rise in temperature. Temperature is raised still further by an increase in voltage, the heat being due simply to the passage of electricity through the water in accordance with the resistance it offers. Although, therefore, d.c. can certainly heat the water, the immense volume of gas liberated

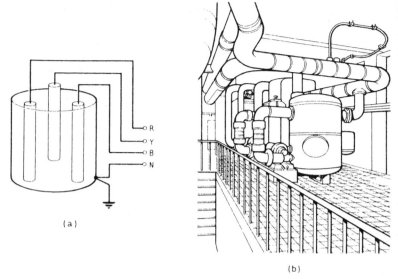

Fig. 16.15. Electrode boiler. (*a*) Diagrammatic view of electrode boiler fed from a 3-phase supply. (*b*) Plant room at English Electric House, showing the two 1200 kW 11 000 V electrode hot-water boilers (Bastian and Allen).

from the water would make it quite impracticable. On the other hand, alternating current when passed through water also heats the water but with hardly any evolution of gas.

In practice, the water in electrode boilers is usually heated by means of 3-phase a.c., three electrodes being placed in the water with the tank solidly earthed and connected to the neutral, as shown diagrammatically in Fig. 16.15(*a*).

There is a similarity in many respects, to a star-connected wire resistance with its neutral point earthed. What, may one ask, is simpler than dipping supply leads into the liquid and obtaining hot water? In practice there are many necessary refinements all hidden by the external appearance of a typical electrode boiler as illustrated in Fig. 16.15(*b*). A twin unit is common to allow for maintenance, repair and standby in the event of a breakdown. Heating by electrodes is essentially for large-scale installations. In fact, the Berkeley Nuclear Laboratories of the Central Electricity Generating Board of Great Britain have a plant of 4 MW capacity.

There are also statutory regulations governing these installations, such as the Electricity Supply Regulations. Most of the requirements of these regulations are met by the I.E.E. Wiring Regulations. They may be summarised as follows:

A circuit-breaker is essential; it must be of a type to make and break all poles simultaneously and have an over-current protective device in each conductor feeding an electrode. The circuit breaker shall be in a position for easy operation and, if not adjacent to boiler, there shall be ready means at hand to open the breaker instantaneously. Where the control device is remote from the boiler, indicating lamps shall also be provided with a clear view to indicate whether the electrode boiler is in an 'ON' or 'OFF' position.

An interlock or similar device must be installed to prevent the electrodes from being alive when work is undertaken on the boiler.

The shell of the boiler to be bonded to the sheath and armour of the incoming cable. The rating of the earthing lead shall not be less than that of the largest conductor. (Note that this contrasts with normal earth-conductor sizes which need not be more than half the current rating of the largest installation wire.) When operated by an earth-leakage circuit-breaker the earthing lead must have a rating not less than the current required by the operating coil, subject to a minimum size of 2·5 mm^2.

Where the supply is in the h.v. range a differential-current e.l.c.b. shall be installed to disconnect the supply from the electrodes in the event of a sustained earth-leakage current in excess of 10% of the rated current of the electrode boiler under normal conditions of operation; 15% may be permitted if it is essential to ensure stability of the boiler. A time delay may be incorporated to prevent unnecessary operation because of short-time unbalance.

In addition to the Electricity Board, the agreement of the water supply authority will be required, and the fact that the heating element is in contact with the water should be brought to their notice, before a boiler or heater of this type is installed. Because of the possibility of stray earth currents affecting GPO lines, the Post Office must be informed within seven days, specifying the position at every point at which the system is connected to earth.

EXERCISES

1. (a) In connecting an electrode boiler installation to a medium-voltage supply, what are the requirements: (i) where the supply is 3-phase, (ii) where the supply is single phase?

(b) Discuss the requirements for the satisfactory performance of an immersion heater installed in a domestic hot-water cylinder, with special reference to maintenance and economy in use. [C]

2. Make a wiring diagram for the following equipment and controls:

Water heating load of 300 kW connected to a 415-V 3-phase 4-wire supply. (The load equally balanced over the three phases.)

Main contactor with 240 V actuating coil.

Time switch (synchronous motor driven).

Thermostat.

Over-pressure switch.

Master hand control to over-ride time switch and thermostat, but *not* pressure switch.

Indicate the current rating of all cables. Note: Apparatus may be indicated diagrammatically or in outline. [C]

3. (a) Compare briefly, *two* different methods of heating a room by electrical means, *excluding the method used below*.

(b) Low-temperature heating panels, with a total load of 10 kW at 240-V, are to be used to heat a room 10 m by 8 m by 4 m. Of the heat generated 50 per cent will be used to heat the air in the room.

Calculate the average temperature rise of the air, if there are two complete changes of air per hour. Density of air is 1·2 kg/m^3. Take specific heat of air as equal to 1010 J/kg°C. [C as modified]

4. (a) Describe two different types of electric water heaters, and indicate the applications for which they are respectively most suited.

(b) Make a neat diagram of the pipe work associated with a pressure-type water heater on the ground floor of a two-storey house. Cold water supply is from a storage tank in the roof. Hot water outlets in kitchen on ground floor and bathroom on first floor.

5. A domestic water-heater of 8 litres capacity is rated at 750 W. Assuming an overall efficiency of 94 per cent, calculate the time required to raise the temperature of the water from 30°C to 94·5°C. Take specific heat of water as 4200 J/kg°C.

Sketch and explain the working of *one* form of thermostat used either in the control of this water heater or for any other heating purpose.

[C as modified]

6. Describe a method of heating a small engineering works by electrical means. Discuss the various types of heaters available and the factors which would affect the power required to heat the workshop. [C]

7. A works reception office 12 m × 10 m with a ceiling height 4 m has a door area of 7 m^2 and a window area of 25 m^2. The office is heated electrically to maintain an average inside temperature of 18°C when the outside temperature is 0°C. Assuming there are two complete changes of air per hour, calculate the kW rating of the heaters.

Heat Transmission Coefficients	(U values) W/m^2°C
Walls—brick and plaster	1·7
Floor—timber on concrete	0·8

Ceiling—plaster   1·2
Doors—wood   2·8
Windows—double glazing   2·9
Specific capacity of air   1010 J/kg°C
Density of air   1·292 kg/m³

Chapter 17

COMMUNICATION SYSTEMS AND EQUIPMENT

17.1 Storage Batteries

In the primary cell, electricity is generated by chemical action. The cell has to be discarded when the active materials are used up. This contrasts with the storage battery (consisting of many secondary cells), where the chemical constituents can be re-activated by passing a direct current in the opposite direction to the discharge.

Storage batteries are used to maintain essential supplies in case of mains failures. The batteries fall into two main classes; the lead-acid and the alkaline types. The voltage given out by each cell is small—2 V approximately for the lead-acid and 1·2 V for alkaline.

Batteries are normally formed by connecting the cells in series. Batteries give the maximum output current when the external connected resistance is equal to the total internal resistance of the cells. This may require a series-parallel arrangement.

For maximum current,

$$n = \sqrt{\frac{NR}{r}}$$

where n = number of cells in each row in series
   N = total number of cells
   R = external resistance
   r = internal resistance of each cell

Example 17.1

A battery consists of twenty-four cells each having an internal resistance of 4 Ω and a voltage of 1 V. The external resistance is 6 Ω. State how the cells are to be connected for maximum current production.

Calculate the current flowing and the p.d. at the battery terminals.

348

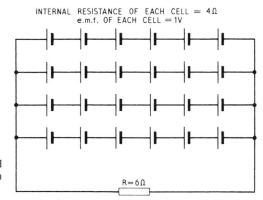

INTERNAL RESISTANCE OF EACH CELL = 4Ω
e.m.f. OF EACH CELL = 1V

R = 6Ω

Fig. 17.1. Series-parallel arrangement for maximum current.

For maximum current,

$$\text{Number of cells in each row in series} = \sqrt{\dfrac{NR}{r}}$$

$$= \sqrt{\dfrac{24 \times 6}{4}}$$

$$= 6$$

There are thus four parallel groups, each group consisting of six cells in series (Fig. 17.1).

The overall battery voltage is equal to six cells in series $= 6\,V$

$$\text{The total internal resistance} = \frac{\text{resistance of one row}}{\text{no. of rows in parallel}}$$

$$= \frac{6 \times 4}{4} = 6\,\Omega$$

$$\text{Total resistance} = 6 + 6 = 12\,\Omega$$

Current flowing $\qquad I = \dfrac{6\,V}{12\,\Omega} \qquad\qquad = \underline{0\cdot5\,A}$

$$\text{Terminal voltage} = E - (I \times \text{total internal resistance})$$

$$= 6 - 0\cdot5 \times 6 \qquad\qquad = \underline{3\,V}$$

Installation and maintenance should follow maker's instructions. In general, batteries are required to be placed in clean, dry and well-ventilated conditions.

17.2. Lead-Acid Cells

The action depends upon the principle that when lead peroxide (PbO_2) and spongy lead (Pb) are immersed in a dilute sulphuric acid (H_2SO_4) electrolyte, a current will flow if connexion is made to the terminals. In the charged condition the chemical composition of the cell is $Pb + PbO_2 + H_2SO_4$. During discharge, portions of the lead peroxide and the spongy lead turn to lead sulphate and water, thus: $2PbSO_4 + H_2O$. The dilution of the acid lowers the specific gravity of the electrolyte. When charging takes place the specific gravity rises and gassing occurs as the water is split up into its constituent parts, hydrogen and oxygen. Any electrolyte lost must be made up by the addition of distilled water. Typical specific gravities of charged and discharged cells are 1·210 and 1·180 respectively. In fully-charged cells, the positive plates are rich chocolate in colour, while the negative plates are slate grey.

The characteristics are given by the charge and discharge curves (Fig. 17.2). Capacity of a cell or battery is stated in ampere-hours for a ten-hour rate of discharge. For example a battery which will supply 50 Ah at the 10-hour rate will give its output at a rate of 5 A over a period of 10 hours.

17.3. Alkaline Cells

The voltage per cell is much less than the lead-acid type and its cost is greater. On the other hand, it has the advantage of robustness due to the stronger plate construction and steel casing. Charges are held for very long periods and the cells can undergo severe charging and discharging without damage.

Alkaline refers to the electrolyte which is mainly a solution of pure *potassium hydroxide* in distilled water. In the nickel-cadmium type (Fig. 17.3(*a*)), positive and negative plates are of similar construction. The plates are essentially pockets made from finely perforated steel strips. Nickel hydrate is used for the positive plate with a conducting

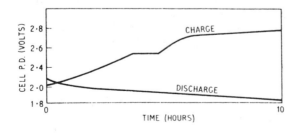

Fig. 17.2. Lead-acid cell characteristics.

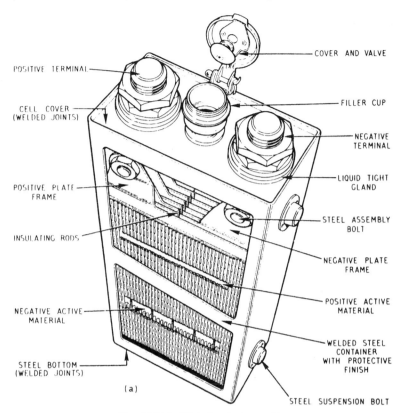

POSITIVE TERMINAL

CELL COVER
(WELDED JOINTS)

POSITIVE PLATE
FRAME

INSULATING RODS

NEGATIVE ACTIVE
MATERIAL

STEEL BOTTOM
(WELDED JOINTS)

COVER AND VALVE

FILLER CUP

NEGATIVE
TERMINAL

LIQUID TIGHT
GLAND

STEEL ASSEMBLY
BOLT

NEGATIVE PLATE
FRAME

POSITIVE ACTIVE
MATERIAL

WELDED STEEL
CONTAINER
WITH PROTECTIVE
FINISH

STEEL SUSPENSION BOLT

(a)

Fig. 17.3(a). Nickel-cadmium cell.

admixture of pure graphite. The negative plate consists of cadmium
oxide with an admixture of a special oxide of iron. The chemical
reaction of charging and discharging transfers oxygen from one
set of plates to the other without affecting the chemical composition
of the electrolyte. In a fully charged battery the nickel hydrate is at
a high degree of oxidation and the negative material is reduced to
pure cadmium. On discharge, the nickel hydrate is reduced to a
lower degree of oxidation and the cadmium plate is oxidised.

Typical charge and discharge curves are shown in Fig. 17.3(b).
During discharge the average voltage is about 1·2 V.

17.4. Charging Methods

The basic circuit is given in Fig. 17.4. The d.c. supply must be
higher than the open-circuit voltage of the battery which is 2·5–2·7 V

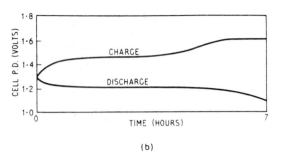

Fig. 17.3(*b*). Nickel-cadmium cell characteristic.

(b)

per cell and approximately 1·6 V per cell for the lead-acid and alkaline types respectively. To prevent the possibility of the current reversing in direction should the battery e.m.f. rise higher than that of the supply, it is usual to fit a reverse current cut-out. The various methods of charging are:

(*a*) CONSTANT CURRENT. As the battery voltage rises during charge, the series resistance in Fig. 17.4, is reduced in order to increase the charging voltage. For charges working from a.c. mains a tapped secondary is often provided on the transformer. The control may be manual or automatic.

(*b*) CONSTANT VOLTAGE. A voltage of 2·3–2·4 V per cell, for lead-acid batteries, is applied directly to the battery. The battery e.m.f. will be low at the beginning of the charge and a heavy current will flow. As the battery e.m.f. increases the current will fall. The circuit resistance prevents too heavy a current at the commencement.

(*c*) TRICKLE CHARGING. This method is of particular importance for standby plant. A scheme for emergency lighting (1500 W, 50 V) for automatic operation is shown in Fig. 17.5. Trickle charging rate is about 1 mA per Ah of capacity for lead-cells up to 100 Ah. The continuous maintenance of this very low current keeps the battery fully charged without any gassing.

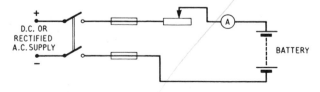

Fig. 17.4. Basic charging circuit.

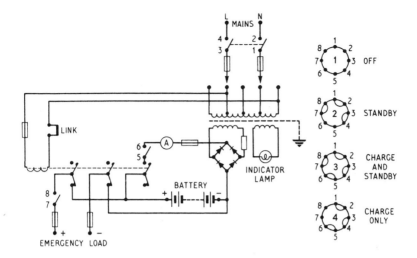

Fig. 17.5. Emergency lighting standby plant.

(*d*) FLOATING BATTERY. Statutory Regulations stipulate that a secondary source of supply for emergency lighting must be available in theatres and cinemas. This has led to the floating system whereby the storage battery, although continuously in circuit, is neither being charged or discharged. The battery is kept fully charged and, in the event of a mains failure, is automatically ready for essential standby services.

Example 17.2

A battery of cells, of total e.m.f. of 40 V and with total internal resistance of 2 Ω, is connected in parallel with a second battery of 44 V and internal resistance 4 Ω. A load resistance of 6 Ω is connected across the ends of the parallel circuit.

Calculate the current in each battery and in the load resistance. Draw a diagram and show the current directions.   [C]

Inserting assumed currents I_1 and I_2 in Fig. 17.6, then by Kirchhoff's voltage law (see also Section 4.6):

MESH ABED

$$40 - 44 = I_1 \times 2 - I_2 \times 4$$
$$- 4 = 2I_1 - 4I_2 \tag{1}$$

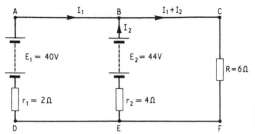

Fig. 17.6. Kirchhoff's laws applied to battery calculations.

MESH BCFE

$$44 = (I_1 + I_2)6 + I_2 \times 4$$

$$44 = 6I_1 + 10I_2 \tag{2}$$

(1) × 3 $\qquad -12 = 6I_1 - 12I_2 \tag{3}$

(2) − (3) $\qquad 56 = 22I_2 \qquad \therefore \quad I_2 = \frac{56}{22} = 2 \cdot 55 \text{ A}$

Insert in (1) $\qquad -4 = 2I_1 - 4 \times 2 \cdot 55 \qquad \therefore \quad I_1 = 3 \cdot 1 \text{ A}$

∴ Current through resistance $R = I_1 + I_2 = 3 \cdot 1 + 2 \cdot 55 = 5 \cdot 65 \text{ A}$

The current directions are as shown in Fig. 17.6.

17.5 Bell Circuits

Figure 17.7 shows the action of a continuous ringing bell. A ledge is fitted to the flat spring and one end of a metal piece rests on the ledge. When the bell push is operated, the metal strip no longer rests on the ledge and rotates so as to meet an auxiliary contact. The bell is now directly across the supply and will continue to ring until the reset cord is pulled. The bell has obvious applications for fire or burglar alarm installations.

Relays are the heart of all communication systems, the essential parts being based on the electromagnetic circuit of the simple trembler bell. In the relay a moving armature is made to open or close contacts, an example being shown in Fig. 17.8 where the operation of the push opens circuit B and closes circuit A. The device can easily be adapted to a number of different switching arrangements, and may contain a latching device to hold on the circuit until released either manually or electrically. Contactors as used for the control of heavy heating loads are also a form of the relay.

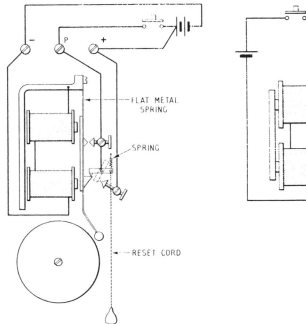

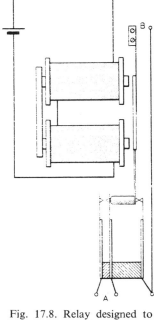

Fig. 17.7. Continuous ringing bell.

Fig. 17.8. Relay designed to make one circuit and break another.

Bell indicator boards are of three main types, namely swinging flag, mechanical replacement and electrical replacement, all operated by a simple form of electromagnet. In each type the mechanism is connected in series with a call bell. The swinging-flag type suffers from the disadvantage that when a call is made the indicator flag only moves for a limited period.

The indicator flag itself can be made to produce a more positive indication by causing a red disc to appear at the window of the indicator board. The disc may be returned to its former position by manually pushing a rod at the side of the board; alternatively, a separate electromagnet may be fitted to produce the same effect when a cancelling bell-push is operated.

Hotel bell communication systems usually include a master indicator board (Fig. 17.9). Tracing out the circuits shows that by pressing any push, in addition to an indication on the board and ringing a bell on the same floor the master indicator records the floor where the call has originated. A relay also operates a bell at

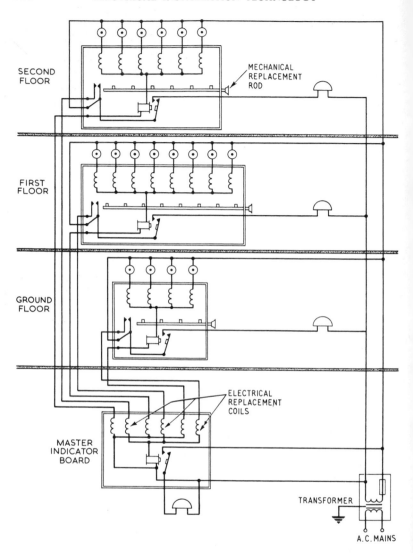

Fig. 17.9. Hotel indicator board system.

the master board position. In this system the mechanical replace-ment rod has a double function, that of manually restoring the indicating disc and, by bridging two contacts, electrically cancelling the indication at the master board.

17.6. Luminous Call Systems

Additional flexibility is achieved through replacing the bells by coloured lamp signals; although audible warnings, such as buzzers, may be incorporated.

In one system the main operation is based on the use of a reset unit, often called re-assurance reset, which holds the call until it is cancelled at the place of origin. The reset houses a small tripping and latching relay. When the caller presses a push-button or pear bell-push, all lamps associated with that call—re-assurance reset, overdoor, corridor or alley-way, section indicator and repeat indicators—will light. At the same time a momentary low-toned audible signal can be made to operate a buzzer in a duty room. All light signals remain on until the attendant presses the caller's reset unit to indicate that the call has been answered. A further feature about the reset unit is that it contains a hollow coloured translucent bell-push button housing a miniature lamp.

A simple application is provided in a manager's or similar executive's office. Outside the door leading to the office is a 2-way light-box indicator for the words 'ENGAGED' and 'ENTER'. When a member of the staff knocks at the door, the manager, by pressing the knob of the reset unit, illuminates the word 'ENTER'. If the manager does not wish to be disturbed, he twists the knob which glows and at the same time lights up the word 'ENGAGED'. Pressing the knob again cancels both signals.

A 3-way indicator with the additional word 'WAIT' gives a further refinement. Four wires are required between the control unit (which here will consist of three knobs) and the indicator (as shown in Fig. 17.10).

Hospitals employ complicated arrangements of light signals which can be operated without disturbing patients. The G.E.C. luminous call system is typical of many possible arrangements (Fig. 17.11).

Fig. 17.10. Wiring for simple luminous call system.

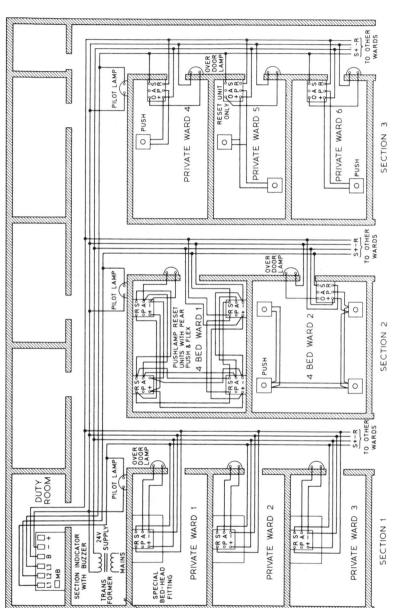

Fig. 17.11. Hospital luminous call system.

In Section 1 a pushlamp reset unit is fitted by each bedside, operating in conjunction with an overdoor lamp unit outside and one common group lamp in the main corridor. In Section 2, multi-bed Ward No. 1 is equipped with a pushlamp reset unit at each bed, operating in conjunction with an overdoor lamp unit outside, but in Ward No. 2 pushes only are fitted beside the beds, working with one common reset unit inside and an overdoor lamp unit outside. Section 3 shows a call push only at each bed with reset unit inside and overdoor lamp outside each ward. It should be noted their

17.7. Internal Telephones

Internal telephone systems are grouped under four main headings:

(1) Single-line telephones designed to work as pairs.
(2) Intercommunication system (push button or key).
(3) Automatic dial system.
(4) Manual switchboards with extensions.

In this book we shall only deal with the first two; the others require more specialist treatment.

Fig. 17.12 gives the circuit for simple 2-way call and speaking which is suitable for short distances. This is one of the many circuits which must be memorised by the student, at least for examination purposes. The task is lightened by familiarisation through tracing out the various circuits, and understanding the working of each part.

With the telephone handset lifted off the cradle and station A button pressed, the bell will ring in station B. By raising the receiver at B both stations can communicate with each other. The battery connexions are reversed so that they are connected in series for the maximum voltage when speaking. Lifting the telephone set and pressing the push of station B causes the bell at station A to ring.

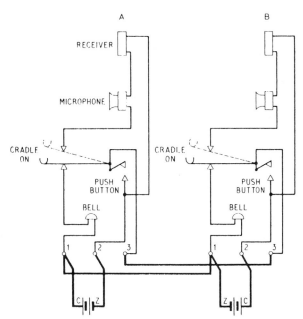

Fig. 17.12. Simple 2-way call and speaking C-positive Z-negative telephones.

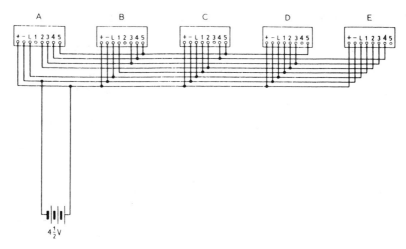

Fig. 17.13. Diagram of connexions for station intercom system.

Intercom systems which are based on the same principle are suitable for 3 to approximately 25 instruments where any station can communicate with any other. Nearly all equipment is of the pushbutton type, and additional facilities offered are from 1 to 4 loudspeaking master telephones. The instruments are fitted with calling buttons in banks of 5, 7 or 10. Any extension of the system involves changing the instruments if they have exceeded their capacity. Intercoms have the facility of immediate connexion to any free line. There is no central equipment other than a small power unit or batteries.

Each intercom telephone is linked by multi-core telephone cables. The number of cores required being one more than the number of lines required to be operated (Fig. 17.13). Except for the wall mounted type, connexion is by a rosette or junction box containing the required number of terminals.

17.8. Fire Alarms

Both the Factory Acts 1961 and the Offices, Shops and Railway Premises Act 1963 stipulate minimum requirements in the case of fire that there shall be *clear* and *audible* warning throughout the building. An earlier statutory requirement (Section 36 (7) of the Factories Act 1937) states 'where in any factory . . . more than 20 persons are employed . . . effective provision shall be made for giving warning in case of fire which shall be clearly audible throughout the building'. The purpose of a fire alarm system is to provide an immediate alarm and so prevent loss of life, also to secure the immediate attention of fire fighting staff. Such a system may consist of suitable bells, bell pushes and a source of supply.

(1) Fire alarm bells must all have a common tone and be distinguishable from other types of bell. Sirens are recommended for overcoming high industrial noise levels. A siren operating at a frequency of 800 Hz gives an audibility range of approximately three-quarters of a mile. Where one bell would not be sufficiently audible throughout the premises served by the system, additional bells, hooters or sirens may become necessary. These devices are generally referred to as 'sounders'.

(2) The pushes or call-points to be of the break glass and press type, painted red. They must be sited in prominent and convenient positions at a height of just under 1·5 m, and installed in the line of escape of the occupants: in particular on the floor landings of staircases and at exits to the street and so located that no person need walk more than 30 m from any position within the premises in order to give alarm.

(3) The supply to the alarm system and cabling should be distinct and separate from the normal mains and wiring. Trickle charged accumulators ensure that the system will operate during a period of mains failure. Consideration must be given to use of heavy cables to offset the effects of voltage drop.

Example 17.3

The British Standard Code of Practice No. CP327, 404/402, 501 for Fire Alarms prescribes that where switches are provided to silence the main alarm sounders, the audible warning must be transferred manually to a supervisory sounder(s). Show by means of a diagram how this can be arranged for a very small building requiring only two break-glass points, and which, therefore, needs no indicator. The circuit is to be so arranged that on the broken glass of the call point affected being replaced, the alarm is to be silenced and the circuit automatically transferred back to the main alarm sounder position should it have been previously switched over to the supervisory sound position. [C & G]

Referring to Fig. 17.14 the alarm contacts are automatically closed by breaking the cover glass plate. The shorting of either of the contacts in this manner rings both alarm bells. The diversion relay is manually operated when the broken glass is required to be replaced. The relay, which is held on by the electromagnet, now

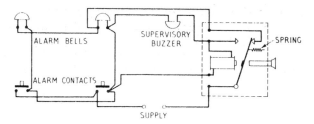

Fig. 17.14. Fire alarm diversion relay with supervisory buzzer.

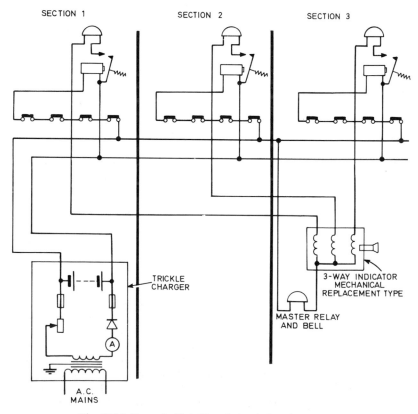

Fig. 17.15. Example 17.4. Closed-circuit fire alarm system.

sounds the supervisory buzzer as a reminder that the alarm bells are no longer in circuit. Replacement of the glass automatically silences the buzzer and restores the system to full working operation. When the call-points are in *parallel*, as in the last example, the arrangement is known as 'open circuit' working. For 'closed circuit' operation the contacts are wired in *series* and operate a relay, to ring the alarm bells, by opening a circuit (Fig. 17.15). The latter method is often preferred since it ensures that the integrity of the wiring is automatically monitored. Clearly any break will immediately set the alarms ringing.

Example 17.4

A closed-circuit fire-alarm system is to be installed in a building which can conveniently be divided into three sections. One alarm and four pushes are required in each section, the bells to ring only in the section where the alarm originated. One main alarm bell and section indicator to be installed in the Fire Post. The system is powered through a battery of secondary cells, circuit pressure 24 V, with a trickle charger connected to the main supply.

Make a diagram of the circuit.   [C]

The diagram is shown in Fig. 17.15.

Automatic fire detectors have the advantage of operating at night or week-ends and provide continuous protection. They are connected as manual call-points. In one type the detector is operated by tilting a mercury tube. A bi-metal strip releases a catch which allows the bottom of the detector, to which the mercury tube is anchored, to tilt and so make or break contact as required.

17.9. Burglar Alarms

Closed circuit working is advisable, as with this system any break or cutting of the wires sets off the alarm bell. A typical example is illustrated in Fig. 17.16. The ball contacts (which may be replaced by magnetic catches) are embedded in door-jambs and window-frames and contacts may also open pressure mats. The 'day switch' allows the system to be cut out while the premises are in normal operation. The 'time delay switch' ensures that a responsible person may leave the premises with the alarm tested and fully working.

Mains supply is rarely adopted due to the possibility of cuts during a critical period. Even trickle charging may not be favoured due to possibility of a breakdown. In this respect manufacturers of burglar alarm equipment recommend dry cells which should be renewed at definite periods.

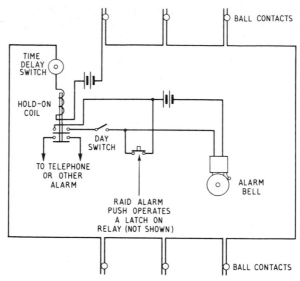

Fig. 17.16. Closed circuit burglar alarm system.

Photo-electric equipment can be adopted to give protection across doorways or access to safes, etc. The breaking of an invisible infra-red beam in conjunction with appropriate relays can be made to set off alarms. Light-sensitive semiconductors may also be employed to have the same effect.

17.10. Codes of Practice

The attention of the reader is drawn to the following:

CP 4166   Burglar, police station terminal.
CP 3116   Fire, heat detection.
CP 5613   Systems for elderly and others at risk.
CP 5345   Explosive atmospheres, apparatus in.
CP 5839   Fire detection and alarm systems in buildings.

EXERCISES

1. Explain the operation of the following battery emergency lighting schemes using rectifiers:
(a) Floating battery system.

(*b*) 'Non-maintained' system with change-over contactor to switch the emergency lights on automatically in the event of interruption of mains supply, and to switch them off on restoration of such.

(*c*) 'Maintained' system with automatic change-over contactor to prevent the emergency lights going out on interruption of mains supply.

Illustrate your answers with diagrams of the basic connexions of system.
[C]

2. Compare a lead-acid cell with alkaline cell, with especial reference to:

(*i*) component parts,

(*ii*) voltage charged and discharged,

(*iii*) charge and discharge characteristics,

(*iv*) physical properties (weight, durability and life),

(*v*) fields of use,

(*vi*) maintenance.   [C]

3. (*a*) What are the constituent parts of a transmitter and receiver in simple telephone instruments.

(*b*) Explain fully the operation of a single transmitter and receiver.   [C]

4. Compare lead-acid and nickel-iron cells in relation to:

(*a*) construction,

(*b*) performance,

(*c*) economics,

(*d*) fields of use.   [C]

5. Draw the full diagram of wiring and internal connexions in the control unit of a fire alarm installation suitable for a three-storey building where each floor is to register separately on a zone indicator, assuming that in the event of an alarm all bells to ring simultaneously. Assume two contacts and two bells per floor. The system is to operate on the closed circuit principle.   [C]

6. Draw any *two* of the following circuits:

(*a*) Busbars, current transformers and metering connexions for a single-phase supply with separate meters for each of lighting, heating and power uses.

(*b*) A battery-operated two station telephone installation, call and reply, both ways.

(*c*) A bell installation comprising a 3-cell battery, 4 pushes, a 4-way indicator with bell, a changeover switch to transfer calls from all pushes to remote bell with continuous ringing relay fitted reset device (indicator inoperative on transferred calls).

(*d*) The internal wiring connexions of a star-delta starter for a 3-phase induction motor.   [C]

Chapter 18

ELECTRONICS

18.1. Thermionic Diode Valve

Whilst valves are being increasingly supplanted by semiconductor devices, very many valves are still in active service and a study of their principles is still demanded by the City & Guilds Electrical Installation courses.

The diode valve consists of an evacuated glass tube containing two electrodes, it acts as a rectifier for converting a.c. to d.c. The principle of action can be understood by reference to Fig. 18.1 where the valve is represented in symbol form. A special oxide coating is given to the cathode filament to enable it to freely emit electrons, the process being accelerated by heating the cathode. Thermionic emission is produced to form the *space charge*. This is the new name given to the electrons which form a cloud round the cathode.

If a positive pole of a battery is connected to the metal anode plate it will attract the negative electrons from the space charge. Current will now flow in a complete circuit made by connecting the negative pole of the battery to the cathode. The current will increase by raising the positive potential at the anode until further increase in electron current is impossible and the valve is said to be saturated (Fig. 18.2).

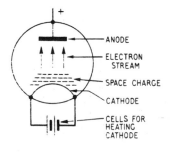

Fig. 18.1. Diode thermionic value.

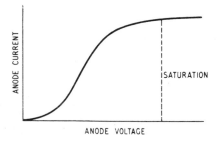

Fig. 18.2. Diode valve characteristic.

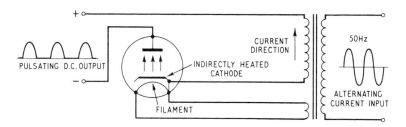

Fig. 18.3. Circuit giving half-wave rectification.

Suppose the battery connexions are reversed so that the anode is connected to the negative pole, then no current will flow, as electrons cannot be emitted from the unheated anode. Thus current through the valve will only flow in one direction.

In a practical circuit (Fig. 18.3) the cathode is indirectly heated. In this way it can be made more robust and can also be heated more conveniently by a.c. A transformer is interposed between the valve and the a.c. supply which also serves the purpose of separating the a.c. and d.c. sides. The heater current being supplied from a separate secondary winding. The electron flow direction, it must be remembered, is opposite to conventional current direction, so that when the anode is made positive with respect to the cathode electrons are attracted to the anode and a current will flow from anode to cathode and leave at the positive terminal.

18.2. The Double Diode

Two anodes in the valve are required for full-wave rectification. The transformer is centre-tapped with a connexion to the cathode (Fig. 18.4). During one $\frac{1}{2}$ cycle, anode A_1 is made positive with

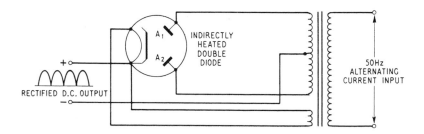

Fig. 18.4. Full-wave rectification with double diode and centre-tapped transformer.

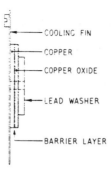

COOLING FIN

COPPER

COPPER OXIDE

LEAD WASHER

BARRIER LAYER

Fig. 18.5. Element of copper-oxide rectifier.

respect to the cathode, and current flows from A_1 to the cathode. When the current reverses during the second $\frac{1}{2}$ cycle, A_2 is now positive with respect to the cathode and current flows from A_2 to the cathode.

The d.c. output while not so fluctuating compared with half-wave rectification is still very uneven. A reservoir capacitor of appropriate value connected across the positive and negative terminals provides some degree of smoothing of the d.c. output. Full smoothing is usually obtained by a π-filter circuit followed by a choke in series and succeeded by another parallel-connected capacitor.

18.3. Metal Rectifiers

Metal rectifiers are much more robust than thermionic valves and require little maintenance. The basic unit for the copper-oxide type is shown in Fig. 18.5.

A copper disc is heat treated on one side to form a thin coating of semiconductor copper oxide. The lead washer makes contact for connexion purposes. Rectification takes place in the barrier layer between copper and the oxide. The unit offers an easy flow of current from the oxide to the copper and a high resistance for connexions made in the reverse direction. The maximum working voltage is about 8 V so that for an increased voltage, units are bolted together in series. Since the maximum current without overheating is 10 A, parallel connexion is required for higher ranges.

Selenium rectifiers, a basic unit of which is shown in Fig. 18.6, have the advantage of permitting a working voltage of some 30 V per unit. A thin layer (about 0·1 mm thick) of the semiconductor selenium is melted on the nickel base plate. The special alloy of

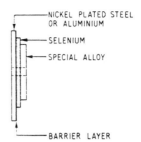

Fig. 18.6. Element of selenium rectifier.

lead, tin, bismuth or cadmium is sprayed on to the selenium. Again the rectification action takes place in the barrier layer. Connexions are made to the base plate and the alloy and the easy-flow current direction is from the plate to the alloy.

Circuits are similar to those shown for the thermionic valve. It is however common to apply the bridge circuit (Fig. 18.7) for full-wave rectification in order to make maximum use of the transformer

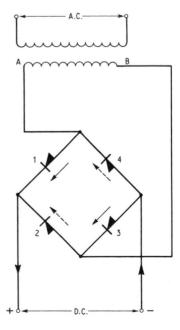

Fig. 18.7. Bridge connexions for full-wave rectification.

windings, as with the centre-tapped method only half of the winding is used at a time. The student can trace out the circuit by considering that during one half of the cycle, point A of the secondary winding will be positive with respect to point B, so that rectifiers 1 and 3 are in operation. During the second half cycle the reverse takes place, bringing rectifiers 2 and 4 in circuit.

18.4. Triode Valves

An additional electrode is fitted between the anode and cathode of the simple diode. This added electrode has an open mesh or spiral construction and is wound with a fine nickel or molybdenum wire. It is taken to an outside connexion and termed the *control grid*, or simply the grid.

The addition of this grid has a critical effect on the operation of the valve. Figure 18.8 shows a simple circuit to produce amplification.

The grid bias battery can be replaced by a cathode resistor (R in Fig. 18.9) which produces automatic biassing. R carries anode current (I_a) and the voltage across it is equal to I_aR. The capacitor C is placed in parallel with R so as to provide a low-impedance path for any varying or a.c. component of the anode current.

We have seen that no current flows in a diode if the anode is made negative with respect to the cathode; thus when no potential is applied to the grid the valve acts as a straight diode. If the grid is now made negative it will repel electrons back to the cathode and so prevent them from reaching the anode. Under these conditions the current will be less than if the grid were at zero potential. The more the grid is made negative the less will be the anode current for a given anode current. If the grid is sufficiently negative, the anode current can be reduced to zero as no electrons will pass through the valve.

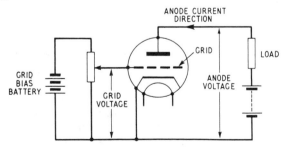

Fig. 18.8. The triode as an amplifier.

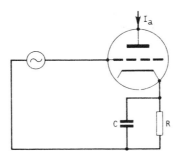

Fig. 18.9. Cathode resistor providing grid bias voltage.

The magnitude of the anode current can therefore be controlled by varying the p.d. between the grid and the cathode. Because the grid is physically nearer to the cathode than is the anode, a small variation in grid voltage produces a larger anode-current change. The valve acts as an amplifier since grid potential changes influence the anode current much more than the same potential changes on the anode. If a change of 2 V on the grid produces a change of 20 V on the anode and a change in anode current of say 5 mA, then there is said to be an amplification factor of $20/2 = 10$.

In radio, signals are applied to the grid in order to produce their amplification in the anode circuit. Gas-filled triodes, which are used for the control of a.c., are known as *thyratrons*.

18.5. Triode Oscillator

Parallel resonance (Section 5.3) may be termed an L/C tuned circuit, see also Fig. 18.10.

Once the capacitor is charged by an external circuit, the supply can be cut off and current will flow in the local tuned circuit. The capacitor will discharge into the inductor and build up its magnetic field. When the capacitor is fully discharged, the field collapses and sets up a back e.m.f. to cause a current which will charge up the capacitor again. This process of current oscillation from capacitor to inductor and vice versa will be repeated until the resistance of the circuit damps down the oscillations.

By connecting the parallel group between the cathode and grid of a triode valve, the anode current will vary in sympathy. If some of the energy from the anode circuit is fed back to the tuned circuit the oscillations can be maintained. Figure 18.11 shows a simple type of feed-back circuit for illustrating this principle. Control of the energy delivered into the L/C circuit is by the mutual inductance

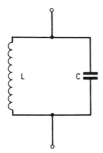

Fig. 18.10. At resonant frequency current will oscillate in the tuned circuit.

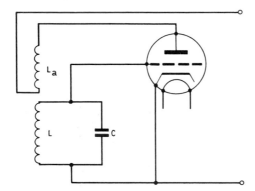

Fig. 18.11. Simple oscillator circuit.

between L_a and L. The entire arrangement is called an oscillator. It acts as a generator of oscillations of varying frequencies depending upon the values of L and C, resonance frequency being equal to $1/2\pi(\sqrt{LC})$.

18.6. The Semiconductor Diode and Transistor

Semiconductor materials, germanium and silicon, may be used for rectification and amplification with the advantage of not requiring a heated cathode and physically much smaller than the corresponding thermionic valve or metal rectifier. Silicon in semiconductors has largely ousted germanium due to its ability to withstand higher temperatures.

Considering germanium, the atoms are arranged in a regular pattern, referred to as a *crystal lattice*. At normal temperatures silicon atoms are held in fixed positions by the interlocking of four electrons in the outer orbit of each atom. *Valence* electrons is

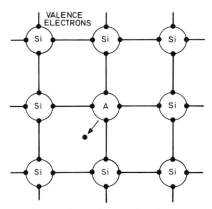

Fig. 18.12. Release of an electron by the introduction of arsenic or antimony into silicon.

the name given to these outer orbital electrons. When minute traces of either arsenic or antimony, each with five valence electrons, are fused with silicon an electron is released (Fig. 18.12). The freed electron becomes a carrier of an electric charge and the treated material turns into *n*-type germanium semiconductor (*n* for negative).

Alternatively, by introducing small amounts of aluminium or indium, which have three valence electrons a converse action takes place. The germanium now has a shortage of electrons so that a 'hole' is left in the lattice which will attract an electron to fill it. The hole left by this electron will be filled by another free electron and thus current will flow. The silicon is now known as *p*-type (*p* for positive).

Joining the two types together forms a silicon junction rectifier (Fig. 18.13).

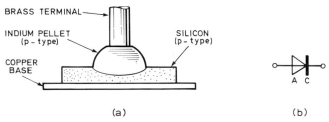

(a)

(b)

Fig. 18.13. (*a*) p–n junction diode. (*b*) Semiconductor *p–n* diode symbol. A: anode, *p*-type; C: cathode, *n*-type.

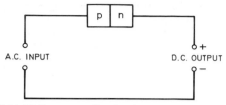

Fig. 18.14. Half-wave rectification by semiconductor diode.

The basic semiconductor rectifier circuit is shown in Fig. 18.14 and gives half-wave rectification. For full-wave rectification a centre-tapped transformer may be employed (Fig. 18.4). The bridge connexion makes full use of the entire transformer secondary winding. To save space the semiconductors are set out horizontally (Fig. 18.15).

To cope with the necessary d.c. voltage, a number of diodes will be required in series. Peak or maximum values (1·414 p.m.s.) also require consideration. The connexions for conversion from 3-phase a.c. to d.c. are given in Fig. 18.16. All these rectified circuits produce a 'raw d.c.' as they contain an a.c. ripple. In order to obtain a smooth d.c. output, the inclusion of a filter circuit as stated at the end of Section 18.2. is necessary.

Transistors consisting of p–n–p or n–p–n types act as triodes and are used where amplification is required. The p–n–p arrangement and the circuit symbol are shown in Fig. 18.17. When a small changing potential is applied between the base and emitter it causes a corresponding variation in the flow of electrons from emitter to collector, thus the collector current forms an amplified replica of the input signal.

In common with other solid-state devices, diodes and transistors are adversely affected by a temperature rise. It is normal procedure therefore to bolt these devices to a mass of metal which will act as a 'heat sink' thereby maintaining the required temperature limit.

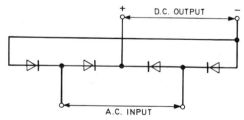

Fig. 18.15. Bridge connexions laid out horizontally.

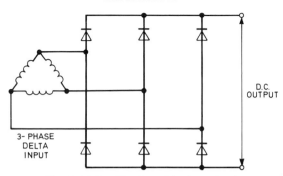

Fig. 18.16. Rectification of 3-phase circuit.

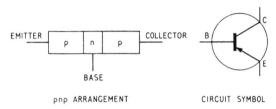

pnp ARRANGEMENT   CIRCUIT SYMBOL

Fig. 18.17. p–n–p transistor.

18.7. Thyristor

Formally known as the silicon-controlled rectifier (SCR), the thyristor has become one of the most important semiconductor devices for the control and switching of small and large alternating currents. From Fig. 18.18 it can be seen that the thyristor consists essentially of four layers of semiconductor material with only three external terminals—anode, cathode and gate, the latter being connected to the inner p layer.

This solid-state device offers a very high resistance to current flowing from anode to cathode until a voltage pulse applied to the

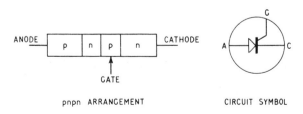

pnpn ARRANGEMENT   CIRCUIT SYMBOL

Fig. 18.18. Thyristor.

gate electrode switches on (fires) the device and allows current to flow from anode to cathode. The arrangement may be employed for the rectification or switch control of small or large currents with the advantage of minimal energy consumption. In addition to the ease with which it can effect a.c. to d.c. rectification, the versatility of the thyristor lends itself to increasing use for:

(1) Speed control of both a.c. and d.c. motors.
(2) Lamp dimming for lighting control.
(3) Inversion, i.e. conversion from d.c. to a.c.

Diacs and triacs consist of multilayer semiconductors with a gate variation of thyristors, allowing current to flow both ways—from anode to cathode, or cathode to anode.

It is, however, important that thyristor circuits also include effective suppressors consisting of appropriate inductor and capacitor networks. Without suppression, interference with radio, television or other electronic apparatus may be experienced if connected to the same supply.

18.8. Printed Circuits

Due to the complexity of electronic circuits—of which computers are but one example—there is an increasing trend towards miniaturisation of both components and wiring. The latter, in the form of printed circuits (Fig. 18.19), are now widely used in industry.

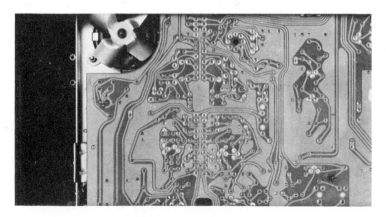

Fig. 18.19. Printed circuit.

Printed circuits consist of a sheet of insulating material, onto which is deposited, by electroplating or other means, a film of copper. The copper film is also deposited onto the inside of eyelet holes, allowing other components to be mounted on the board. The pattern of the required wiring is then printed with acid resisting ink onto the copper film. Then the untreated areas of copper can be etched away in an acid bath, leaving the required conductor pattern.

18.9. Cathode Ray Oscilloscope

The instrument is constructed on the lines of the familiar television tube. Here it is used to show a.c. waveforms although the display can be calibrated for use as a voltmeter and many other purposes.

Action depends upon a beam of electrons issuing from the heated cathode (Fig. 18.20) which is made to strike the fluorescent screen and produce a light spot. This beam is deflected by a p.d. between the plates to trace out the waveform. Essential requirements for the operation are:

(1) An anode A, which is at a high positive potential and attracts the negative electrons, while a second anode A_2 further accelerates the electron movement towards the screen.

(2) The grid electrode is at a potential negative to the anodes and serves to focus the electron stream.

(3) Varying potentials at the X and Y plates produce vertical or horizontal movements of the electron beam. By a combination of deflections the beam can be made to trace out any desired waveform.

(4) For continuous production of the wave on the screen an X plate horizontal deflection is essential for supplying the time scale, known as the *time base*, otherwise the spot would simply be drawn into a line. The wave pattern as maintained by the X p.d. itself

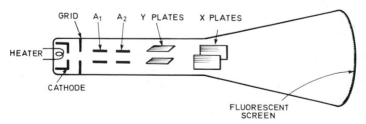

Fig. 18.20. Cathode ray oscilloscope.

V

TIME

Fig. 18.21. Saw-tooth wave producing time base.

follows a saw-tooth wave pattern (Fig. 18.21). This waveform has the property of providing a voltage which increases uniformly with time to reach a position where it falls sharply and then repeats the sequence.

EXERCISES

1. Show by means of diagrams the meaning of (a) half-wave rectification, (b) full-wave rectification.
Describe, with sketches, one form of full-wave rectifier, and explain clearly its action.

2. (a) Describe a metal-plate rectifier and explain how a unidirectional current is produced.
(b) Draw the diagram of a single-phase bridge circuit, marking the positive and negative output terminals.
(c) What are the I.E.E. Regulations covering the installation of these rectifiers?

3. Define the following terms: (a) impedance, (b) transistor, (c) solidly-earthed, (d) frequency, (e) power-factor, (f) transformer, (g) double-insulation, (h) saturation (magnetic).

4. Describe with the aid of a diagram and sketches:
(a) A full-wave metal rectifier suitable for supplying electrical energy to a direct-current control circuit from a single-phase alternating-current supply.
(b) A rectifier suitable for installation in a factory to supply a large group of variable-speed direct-current motors from a 415-V, 3-phase alternating-current supply.
Explain briefly how each rectifier works, and draw a diagram of internal connexions and leads from the supply and to the load in each case.

5. Describe briefly three methods of producing direct current from an alternating current supply. [C]

6. (a) Describe how a triode valve can be used as an oscillator.
(b) Explain which circuit components determine the frequency of oscillation.
Describe with sketches, a single phase full-wave rectifier of the glass-bulb type and explain its operation.

7. Explain (i) operation of rectification in the diode valve, (ii) operation of amplification in the triode valve.

8. Describe the action of the p–n semiconductor diode.

Chapter 19

MANAGEMENT

19.1. Site Management

Efficient management and supervision are essential to a successful electrical installation business. In an industry with such a high labour cost content—on many sites reaching as much as 60% of the total contract costs—an understanding of labour relations is essential. There must also be a real appreciation of the commercial issues involved and the ability to take appropriate action, without hesitation, at each stage of the work.

The type of contract and site conditions considerably influence electrical installation management procedures. Generally the main contract is awarded as a result of competitive tendering, and there are a number of subcontractors for the various works to be undertaken, the electrical contractor being held responsible for part, at least, of the engineering services.

All subcontractors have specialist contracts so that coordination is one of the keys to a successful project. It is one of the important duties of the electrical technician and planner to ensure that his installation fits exactly into this jigsaw pattern. The timing for installing the intake work, rising and lateral mains, wiring and accessories must be right, otherwise delays and consequent financial losses are bound to occur. That the contracting staff forms part of a team is perhaps even more clearly brought out by the current trend towards integrated services—comprising heating, lighting, air-conditioning and sound insulation in combined units. In certain instances the combination of integrated services may require control by a consortium of firms.

Clearly no two projects are alike, and, in fact, a distinguishing feature of electrical installation engineering is the great variety of its work. Yet, whatever the size or complexity, the supervisory personnel must have certain basic objectives of costing, planning and site and office control with the prime aim of increasing productivity and efficiency.

19.2. Specification

In addition to contract conditions, a specification may be regarded as a complete description of the work to be carried out. There is a wide variation between types of specification, from the instructions to wire a few lighting points and socket-outlets, to a 100 or more page booklet containing precise and detailed clauses as required for a complex and important undertaking. Again some are written in simple language while others are full of technical and semi-legal terminology and are probably accompanied by a multitude of site drawings. For the larger contracts, diagrams of busbar, armoured cable and conduit runs may be included together with the layout of all control gear and special instructions relating to the particular building.

Since specifications form the basis of a contract between contractor and client, and perhaps other parties, the wording must be clearly understood. Where there is any doubt, *written confirmation* of the exact meaning should be obtained.

When the work is actually in progress, continuous checks must be carried out. This is necessary in order to ensure that the installation as fitted complies in every respect with the instructions as set out in the specification. Table 19.1 illustrates one page of a typical specification drawn up for a contract as required for a factory installation. The extract gives an indication of the responsibilities of personnel in charge of the electrical installation and sets out the normal 12 months' guarantee.

19.3. Elements of Estimating

Electrical installation work is a highly competitive business, and contracts are usually given to firms that submit the lowest tender. Therefore one of the major factors in operating a successful contracting business is accuracy in compiling quotations. If prices are too high a contractor will secure no order; if too low they will certainly land him in trouble. Although correct estimating is vitally important, it is, of course, only part of running a business.

As a guide there is a Standard Form of Accounts available, and an Accounting Procedure has been produced. Included in these documents are specimen Trading Accounts, Profit and Loss Accounts and Balance Sheets. Model cost cards, a suggested form of bookkeeping, estimating and pricing policies, monthly budgeting and a system of inter-firm comparison are also described.

Accurate pricing requires the estimator to have a sound technical and practical background in addition to a keen business acumen.

TABLE 19.1. SPECIFICATION EXTRACT 381

10. *Labour and Materials* (*Contd.*)

All materials and articles used in the works shall be of the highest quality and execution. They must also comply to the relevant British Standards.

All workmanship and materials will be carried out to the entire satisfaction of the Building Surveyors or their appointed Electrical Consultant.

11. *Site visits and programme*

The main building works are in progress and the Contractor is to satisfy himself as to the character of the work to be carried out and the conditions effecting the execution of the sub-contract generally. The Contractor is also expected to visit the Alpha Printing Works of Messrs. John King & Co. Ltd., and to be familiar with the electrical connection details of the machinery and equipment which is to be reinstalled at the Central Street site.

The building works are scheduled for completion by ...... and due allowance should be made for attending at the site for intermittent periods as may be required by the Builder for the purpose of marking out, setting the position of concealed conduit and other incidental works which may be required on site prior to commencing the sub-contract works.

12. *Regulations*

The Contractor shall ascertain and act in accordance with any statutory regulations, local byelaws and regulations which govern the installation of electrical equipment in Factory premises.

All wiring and apparatus shall satisfy the requirements of the appropriate British Standard Specifications, and will comply in all respects with the recommendations prescribed by the Institution of Electrical Engineers, Electricity Supply Regulations and the Health and Safety at Work Act.

13. *Liability for defects*

The Contractor is to allow for making good within a reasonable time after receipt of written instructions from the Building Surveyors any defects of any kind which may arise within a period of TWELVE MONTHS from the date of acceptance of the completed installation by the Building Surveyor and which may be due to materials and workmanship not in accordance with this sub-contract and are not caused through accident, misuse or neglect.

In this connection the Contractor must clearly understand that his liability for defects includes any item or article supplied by merchants or other specialist manufacturer or supplier.

The Contractor must therefore obtain any indemnification he deems desirable from any merchant, manufacturer or supplier before he places his order for goods.

14. *Builders' work*

All builders' work incidental to the installation will be carried out by the Builder.

The Contractor is to supply to the Builder details and dimensions of any cutting away and the Contractor will be responsible

He must be able to visualise the conditions under which the work will be done. Exposure to weather and arduous circumstances may have to be considered.

CONDITIONS. Clearly, electrical contracting is vastly different from work in a factory where, under controlled conditions, operations can be timed and costs calculated to an exact figure.

No such precision is possible—nor normally desirable—with wiring jobs where full supervision is difficult and the work may be subject to the hazards of the building industry. Installation operations often have to dovetail into the work of other trades, over which the electrical contractor has no control. The work may be subject to hold-ups, or to sudden rushes, which can entail expensive overtime costs. A smooth flow of operations, while desirable, is rarely attained.

Every clause and even individual words of the specification must be studied, checked and rechecked. A missing or mistaken letter or figure of a catalogue number can make a difference of hundreds of pounds. Not only what is *in* the specification but also what is *left out* may be of vital importance. Items such as 'rise and fall' in a period of rapid inflation and 'cutting away and making good' may easily make the difference between profit and loss.

Visiting the site or job is a must. Where an existing building is concerned, full details should be obtained. These include, for example, the type of fixings (invariably a time-consuming process) and as to whether the partitions and walls lend themselves to easy concealment of cables. Parking facilities should also be noted. Any special difficulties in carrying out the actual wiring must be taken into account. On the credit side an experienced eye will spot where savings can be effected, such as by the re-use of existing wireways.

19.4. Pricing

For the estimate, it is necessary to take off accurate quantities from the drawings. Where the runs are repetitive or nearly repetitive an average figure is often used. Checks, with care, should be made against cost records of previous contracts. For the estimate proper there are four major items; namely materials, labour, overheads and profit.

The estimating engineer must be in possession of all the appropriate catalogues, which will be required to be continually kept up-to-date. As a safeguard it is advisable, in the absence of a 'rise and fall' clause, to add a covering statement that the estimate is based on the prices ruling at the date when the tender is despatched.

Prices *in writing* should be obtained for the larger pieces of equipment.

Except for pre-assembled units, allowance must be made for waste, usually in the order of 10% in cutting conduit and 5% in cable. Provision must also be made for unexpected diversion of runs, as they may bear little relation to the straight lines on the drawings.

19.5. Labour

Labour is the most difficult and variable item to assess. It is rare for two people to agree on the time needed to complete an operation. This is not surprising when it is affected by so many factors. In addition to rates of pay, such factors as interest, skill, loyalty, physical fitness, site condition and organisation, technical training, cooperation of the main contractors, weather and even the home life of the operator may all have a bearing.

Only experience and records can prevent the assessment from becoming a matter of chance. It is precisely here that full and correct costing may reap rewards. A discerning analysis of time sheets, to give the desired information, often produces valuable results. The estimator is then able to use this information for obtaining labour values by the system of *man-hours constants*. These give the average times taken to complete given operations under normal conditions, and they permit a reasonably systematic approach to estimating. The labour constants must be built up. They are not static and require to be constantly revised. These constants include factors such as preparation, marking out, collecting and packing tools.

19.6. On-costs, Provisional and Prime Cost Sums

Charges for labour must allow for statutory insurance, pension, holidays with pay and Value Added Tax. All costs connected with labour which cannot be conveniently split up into any particular contract are known as 'on-costs'. They include supervisor's salary and expenses, transport costs, repair and maintenance of plant. When allocated to an installation, on-costs should be converted to a percentage.

OVERHEAD CHARGES. Sometimes known as establishment charges, these have to include general office expenses, rent, rates, lighting and heating, depreciation of furniture and office equipment. Overheads should be ascertained annually and a percentage figure obtained for inclusion in the estimate.

A PROVISIONAL SUM. This is the amount which the client is prepared to spend on certain unnamed equipment on works to be specified at a later date. The sum includes the normal trade discount and, similarly to on- and overhead costs, is simply added to the gross value of the completed contract.

Any *prime cost* (often abbreviated to PC) item should be treated in the same way as the cost of any of the materials. A percentage profit being added to PC items.

19.7. Bill of Quantities and Daywork

For local government work there is an increasing tendency to employ this form of estimating the cost of new installations. Here the complete contract is split into a number of individual items and for costing purposes the contractor is issued with a comprehensive list. The Bill will also probably include schedule rates, which the estimator must complete, for labour.

Bills of quantities are designed to reduce the element of chance in tendering. They may act in favour of the contractor, who might otherwise miss out essential materials, equipment or other sections of the contract.

Daywork is a 'cost-plus' method for contracts where it may be difficult—owing to unforeseen circumstances—to forecast or assess the cost of the proposed work. Owing to the possibility of abuse in calculating the sum to be charged to the client, safeguards have been set out by the Royal Institution of Chartered Surveyors. The basis of agreement is that the percentage cost shall be based on the hours actually worked.

19.8. Bar Charts

Simple pictorial methods of displaying changes in project developments greatly facilitate work progress. They draw attention to weaknesses so that appropriate remedial action can be taken. As can be seen from Fig. 19.1, the bar chart consists of a series of rectangles, with the lengths varying according to the quantity to be measured. Here the overheads show a rather sharp rise towards the end of the year. The chart also has a positive aspect, as it may reveal unexpected savings which also demand investigation and result in improved efficiency.

Furthermore, bar charts lend themselves to a variety of forms. Figure 19.2 shows an application to the control of the wiring of an office block. At a glance it can be seen that the top two floors are completely tubed and wired, in addition to the fifth floor being

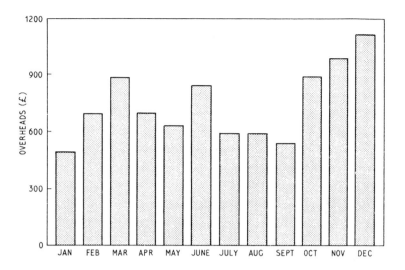

Fig. 19.1. Bar chart showing monthly changes in overhead costs.

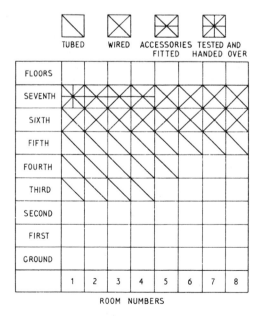

Fig. 19.2. Chart showing progress in wiring of an office block.

tubed only. Also tubed, are rooms one to five on the fourth floor and rooms one to four on the third floor. On the seventh floor, rooms one to four have accessories fitted, while room one has been tested and handed over to the main contractor or client.

These charts can easily be adapted to industrial and other installations.

19.9. Critical Path Analysis (C.P.A.)

This type of network analysis technique is a comparatively recent development and is designed to show how any given series of tasks, involving time elements, can be completed in the shortest possible time. The method is particularly suitable for management who are responsible for carrying through electrical installation projects. It is fast becoming a standard system whenever a 'flow' of operations becomes necessary for the completion of a scheme within a certain time.

The building industry is not normally amenable to time and motion study and in fact has been accused of being slow in adopting modern industrial methods. It is therefore encouraging to see critical path analysis in use by civil engineering contractors. C.P.A. is not only suitable for the larger sites; smaller contracts can also benefit from its use.

As in any project, there are three essentials, namely planning, scheduling and controlling. These are, of course, normal aspects of management. What is new is the graphical method of presentation. This is made up of circles and arrowed lines (Fig. 19.3) which are

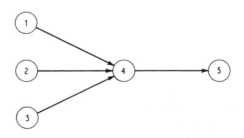

Fig. 19.3. Representation of activities.

part and parcel of the analysis. To see how it works, critical path analysis in a simplified form is applied to a medium-sized industrial building. The contract concerns an addition to an existing installation as follows:

  I   34–80 W luminaires
 II   3—Twin 80 W luminaires
III   5—Tungsten lamp points
 IV   7—150 W single phase motors
  V   18—15 A triple pole socket-outlets
 VI   Re-arrange existing and install new intake switchgear

Wiring system: mineral insulated copper sheath cable.

PLANNING. The entire work is first split up into a number of basic elements. The elements, which are numbered, are called *events:*

Event 1   Commence work on site
Event 2   Mark and set out points
Event 3   Re-arrange switchgear at intake
Event 4   Fix new switchgear
Event 5   Make metal fixing straps
Event 6   Fix fluorescent fittings
Event 7   Fix wiring clips
Event 8   Wire points
Event 9   Connect fittings and motors
Event 10 Connect intake
Event 11 Test

SCHEDULING. Lines with arrowheads may be made to represent *activities.* A time must now be allotted to each activity, which could be as follows:

		Hours
Activity 1–2	Mark out points	4
Activity 1–3	Re-arrange switchgear	8
Activity 3–4	Fix switchgear	14
Activity 2–5	Make fixing straps	4
Activity 2–6	Fix fluorescent fittings	20
Activity 5–7	Fix straps	4
Activity 6–8	Wire all points	60
Activity 8–9	Connect points	16
Activity 4–10	Complete intake connections	6
Activity 9–11	Test	2

The network may also entail the use of *dummy* activities of zero duration and is shown by broken lines. Figure 19.4 indicates the

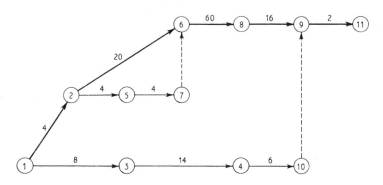

Fig. 19.4. Network complete with activity times (hours) thick lines represent the *critical path*.

completed network with activity times. It will be observed that dummy lines also act as restraints. The straps must obviously be fixed (activities 5–7) before the wiring operation (activities 6–8). In addition, the intake requires to be connected (activities 4–10) before testing (activities 9–11).

CONTROL. There are three paths from commencement on site (1) until completion (11), the total times for each of these being:

$$4+20+60+16+2 = 102 \text{ h}$$

$$4+4+4+60+16+2 = 90 \text{ h}$$

$$8+14+6+2 = 30 \text{ h}$$

The path with the longest time, i.e. 102 hours, represents paradoxically, the shortest possible time for the completion of the contract. It is this route which is called the critical path. Other paths have what is known as a certain amount of 'float', and if spread out will not delay the finish of the contract provided the total time taken to complete the path does not exceed the 102 hours.

Critical path networks enable the supervisory staff to appreciate precisely the manner in which projects are proceeding. Thus the graphical representation enables control to be effected so that, if any activity exceeds its allotted time, remedial action can be rapidly carried out. The various links which make the installation can also be easily observed.

19.10. Earliest and Latest Times

For a deeper analysis the circles or nodes may be separated by a quadrant into four segments (Fig. 19.5.)

Fig. 19.5. Network quadrants. E.E.T., earliest event time; L.E.T., latest event time; E.N., event number.

The *earliest event time* (e.e.t.) is the earliest time by which all operations leading towards a node can be completed. The *latest event time* (l.e.t.) is the latest time by which all the operations leading to a node can be completed. *Float* is the difference in time between earliest and latest event times at a particular node. This allows some increase in the use of spare capacity while does not effect the critical path. Critical events are those with the same node earliest and latest times. Linking these nodes comprise the critical path. Again paradoxically the critical path is the path through the network which takes the longest time. The method can be examined by working through a City & Guilds' Question.

Example 19.1

A bar chart for a small project is shown as follows. It must be noted that:

activities D and E can only commence when B is complete;
activity F can commence when A is complete;
activities G and H can only commence when D is complete;
activity I can only commence when E, F and G are complete;
activity C does not restrict other activities.

(a) Prepare a network diagram of the programme. (b) Clearly mark on the diagram the critical path. (c) State how much longer time could be spent on activity E without increasing the total project time.

[C]

ACTIVITY	5	10	15	20	25	30	35	40	45	50	55
A	▬	▬	▬								
B	▬	▬	▬								
C	▬	▬	▬								
D					▬						
E					▬	▬					
F				▬	▬						
G							▬				
H							▬	▬			
I									▬	▬	

WEEKS

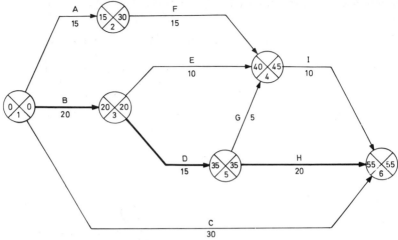

Fig. 19.6. Network arrangement of bar chart (thick lines indicate critical path).

(*a*) Translating the bar chart into an appropriate network is governed by the restraints as given in the question i.e. activity E and B cannot start until B has been completed. Similarly G and H cannot commence until activity D is finished. Activity C forms a separate path as it is not dependent on, nor restricts, any other of the activities. The network is set out in Fig. 19.6. Starting from week 0 (event 1):

$$\begin{aligned}
\text{earliest time for event } 2 &= \;\;0 + 15 &&= 15 \text{ weeks} \\
\text{earliest time for event } 3 &= \;\;0 + 20 &&= 20 \text{ weeks} \\
\text{earliest time for event } 4 &= 20 + 15 + 5 &&= 40 \text{ weeks}
\end{aligned}$$

(there are three paths to event 4, but the one with the longest time is taken)

$$\text{earliest time for event } 5 = 20 + 15 \qquad = 35 \text{ weeks}$$

earliest time for completion (event 6)

$$= 35 + 20 \qquad = 55 \text{ weeks}$$

To determine the latest event times, it is necessary to work *backwards* from the last event (i.e. event 6) and *subtract* activity times.

$$\begin{aligned}
\text{Latest time at event } \;5 &= 55 - 20 &&= 35 \text{ weeks} \\
\text{Latest time for event } 4 &= 55 - 10 &&= 45 \text{ weeks} \\
\text{Latest time for event } 3 &= 35 - 15 &&= 20 \text{ weeks} \\
\text{Latest time for event } 2 &= 45 - 15 &&= 30 \text{ weeks} \\
\text{Latest time for event } 1 &= 20 - 20 &&= 0
\end{aligned}$$

(there are two other chains $30-15$ and $55-30$, but the shortest is taken).

(b) The critical path lies where the earliest and latest event times are equal and is indicated by the thick lines in Fig. 19.6.

(c) The earliest activity time that E can commence is week 20. The latest possible time it can finish is week 45, so that the maximum available time is $44-20=25$ weeks. Account must now be taken of the duration time of 10 weeks, therefore the time that could be spent on activity E without increasing the total project time is $25-10$ i.e. 15 weeks.

19.11. Logic Gates and Circuits

The solutions to the vast complexity of modern industrial control and computer problems are, in essence, obtained by giving *yes* and *no* answers to relevant questions. These are applied to circuits which are either switched *on* or switched *off* (on/off). The required solution is fed from a two-state or binary arrangement. This switching on and off can be understood as a logic equivalent to gates.

Each separate amount of information is represented by a letter to provide equations conforming to Boolean algebra. Statements about the information must be *true* or *false*, which again corresponds to the on/off arrangement.

Conditions when true have a logic value of 1, e.g. the statement 'copper is a conductor' could be represented by the letter A. Since the statement is true, $A=1$. If B corresponds to 'socket-outlets have pins', this is false so that the Boolean equation becomes $B=0$.

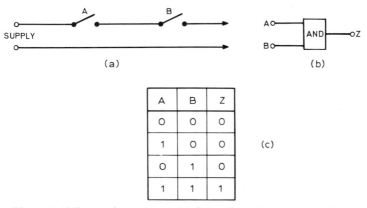

Fig. 19.7. AND equation $A \cdot B = Z$. (a) Gates, (b) symbol, (c) truth table.

AND GATE. In Fig. 9.7(a), A represents a supply switch and B a thermostat for the simple control of a heating system. The switches are in series and both must be in the on position for the heating system to operate. The Boolean equation becomes $A \cdot B = Z$ and reads as 'A and B equals Z'. The logic symbol form is given in Fig. 19.7(b). To display all possible combinations of switching a truth table (Fig. 19.7(c)) is compiled.

With three switch controls in series (Fig. 19.8(a)), the equation becomes $A \cdot B \cdot C = Z$ and reads 'A and B and C equals Z'. Fig. 19.8(b) and 19.8(c) show the circuit in the sumbol form and the truth table respectively.

OR GATE. A circuit comprises two bell pushes connected in parallel for the operation of an alarm bell (Fig. 19.9(a)), where the bell would ring from the closing of either A or B push to form an OR gate. The Boolean equation becomes $A + B = Z$. As a contrast to normal equations, it reads 'A or B equals Z'. Fig. 19.9(b) shows the symbol form and Fig. 19.9(c) the truth table. The circuit, symbol and truth table are given in Fig. 19.10 for $A + B + C = Z$, reading 'A or B or C equals Z'.

It will be noticed that the number of different switch combinations obtained is 2^n, where n stands for the number of switches.

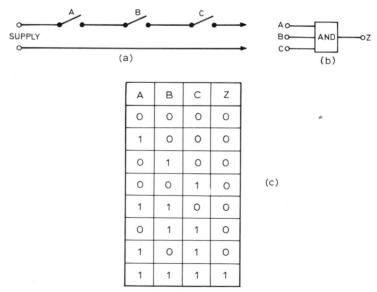

Fig. 19.8. AND equation $A \cdot B \cdot C = Z$. (a) Gates, (b) symbol, (c) truth table.

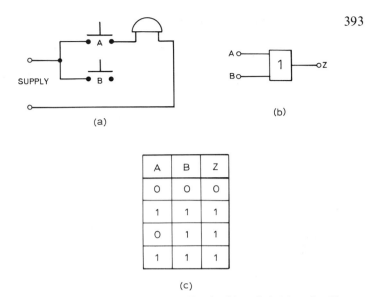

Fig. 19.9. OR equation $A + B = Z$. (a) Circuit, (b) symbol, (c) truth table.

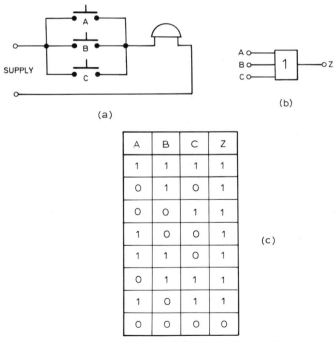

Fig. 19.10. OR equation $A + B + C = Z$. (a) Circuit, (b) symbol, (c) truth table.

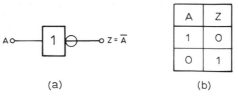

A	Z
1	0
0	1

(a)   (b)

Fig. 19.11. NOT equation $\bar{A} = Z$. (a) Symbol, (b) truth table.

NOT GATE. In a chemical plant a warning signal is to be sounded when the temperature is not below a specified temperature. Here the equation is simply $\bar{A} = Z$ is put to negate or reverse a statement. This is a gate giving 1 for 0 and 0 for 1. This NOT gate reverses the output. The symbol and the truth table for the equation are shown in Fig. 19.11(a) and 19.11(b).

NAND GATE. The three basic gates as described above may be linked. Here the linking of a NOT and an AND gate is put in a shortened form from NOT–AND to NAND. The equation for the two functions becomes $\overline{AB} = \bar{Z} = X$. It is important to note the negating bar above AB and Z and that X is positive. Fig. 9.12 shows the symbol and truth table.

NOR GATE. This is another linkage function consisting of NOT–OR i.e. NOR. The equation for gates A,B,C becomes $\overline{A+B+C} = Z$ and reads 'bar A or B or C equals Z'. The symbol form and truth table is shown in Fig. 19.13.

SWITCHING. On/off switching may be mechanical or electromagnetically operated. For complex situations requiring thousands of switch points, electronic methods are now adopted.

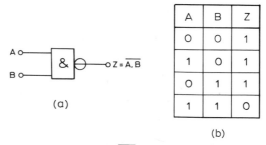

A	B	Z
0	0	1
1	0	1
0	1	1
1	1	0

(a)

(b)

Fig. 19.12. NAND equation $\overline{A \cdot B} = Z$. (a) Symbol, (b) truth table.

(a)

A	B	C	Z
0	0	0	1
0	0	1	0
0	1	0	0
0	1	1	0
1	0	0	0
1	0	1	0
1	1	0	0
1	1	1	0

(b)

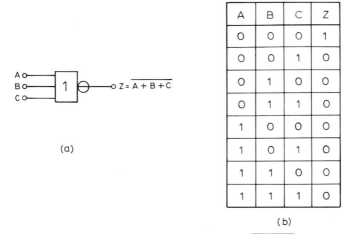

Fig. 19.13. Logic diagram and truth table for $\overline{A+B+C} = Z$.

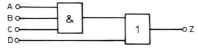

Fig. 19.14. Logic diagram for $A{\cdot}B{\cdot}C+D = Z$.

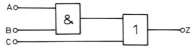

Fig. 19.15. Logic diagram for $A{\cdot}B+C = Z$.

A	B	C	Z
0	0	0	0
1	0	0	0
0	1	0	0
0	1	1	1
1	0	1	1
1	1	0	0
0	0	1	1
1	1	1	1

Fig. 19.16. Truth table.

Solid state devices require little space combining with advantages of speed and efficiency. Semiconductor diodes with half-wave rectification can be used to act as a gate switch.

Transistors commonly function as switches in logic circuits. A changing potential applied to the base and emitter produces an amplified flow of electrons from emitter to cathode to produce the states of 0 and 1.

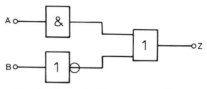

Fig. 19.17. Logic diagram for $A \cdot \bar{B} = Z$.

A	B	$\bar{B}$	C
0	0	1	0
0	1	0	0
1	0	1	1
1	1	0	0

Fig. 19.18. Truth table.

Example 19.2

(a) *Draw a logic diagram for the Boolean equation* $A \cdot B \cdot C + D = Z$.

(b) *Set out a truth table for* $A \cdot B + C = Z$.

(c) *Give a logic diagram and corresponding truth table for* $A \cdot \bar{B} = Z$.

(a) Fig. 19.14 shows the logic diagram.

(b) The logic diagram must first be drawn (Fig. 19.15) for the truth table (Fig. 19.16).

(c) Logic diagram and truth table are given in Fig. 19.17 and 19.18 respectively.

EXERCISES

1. (a) Explain the difference between (i) a lump sum type of contract and (ii) one which is dependent on a bill of quantities.

(b) If the architect issues a revised drawing, explain the procedure which should be adopted to record the variation and ensure correct payment for the variation in each of the types of contract mentioned in (*a*) above, (*i*) if the installation to the superseded drawing had not been commenced and (*ii*) if the work to the superseded drawing was practically completed.

2. (*a*) Explain the action to be taken where a serious electrical accident takes place causing shock and serious burns to the operative.

(*b*) Detail the post-accident reporting procedure.

3. Prepare notes for guidance to be given by a site foreman to a job/shop representative (shop steward) in order to ensure good labour relations on a site. The areas to be covered should include clocking on and off routines, tea and meal breaks, washing and changing facilities, grievance and disputes procedure, meetings of men and redundancy procedure.

4. Detail the items which should be regularly recorded in a site diary.

5. Draw a critical path network for the installation of lighting, block storage, heating and power services in a new house showing such items of other trades' work as are relevant to the electrical installation. Assume the construction is traditional with solid ground floor, timbered first floor and plastered walls.

6. A standby generating set is to be installed in a building in which some reconstruction work is required. The estimated time for each activity involved is shown below.

Activity detail	Time (*days*)
A clearing site prior to laying foundations	1
B completion of site clearing	2
C foundation work	2
D construction of walls	2
E cabling to existing switchgear	3
F installation of set	2
G installation of ancillary apparatus	2
H electrical connections	1
I testing and commissioning	1

The cabling work cannot be started until all the site is cleared and the set cannot be installed until all constructional work has been completed.

(*a*) Draw a network diagram for the programme. (*b*) Determine the minimum time necessary for completion. (*c*) State what effect, if any, there would be on the total programme time if the time taken to carry out the cabling to the existing switchgear were increased to 5 days.

ANSWERS TO EXERCISES

Abbreviations

[C] City and Guilds Electrical Installation Work, Course C
[T] City and Guilds Electrical Technicians Course
[WJ] Welsh Joint Education Committee

Chapter 1

2. (i) 0·628 T, (ii) 132 N
3. 2·5 N
6. 3·84 A
9. 5·1N

Chapter 2

3. (a) 664 V, (b) 0·2 N
5. (d) 100 V
7. (i) 0·2 T, (ii) 0·1 N, (iii) 0·04 J

Chapter 3

1. (b) (i) 3·7 μF ; $V_1 = 111$ V, $V_2 = 74$ V, $V_3 = 55·5$ V, (ii) 36 μF, $V = 240$ V
3. 0·06 J, 120 V, 0·036 J
4. (b) 18 A
5. 10 μC/m^2, 20 kV/m
6. (i) 2 μF
7. (i) 0·2 μC, (ii) 4 μC/m^2, (iii) 200 kV/m, 0·2 mm
10. (b) (i) 2 μF, 100 V, 50 V, (ii) 9 μF, 150 V

Chapter 4

1. 77·5 A, 7·5 A, 42·5 A ; Q, 230·7 V ; R, 229·8 A
2. 6·67 V from B to D. 12 V
3. 2·7 mA from B to D
4. 112·5 Ω
5. 43·4 Ω from A to C
6. (a) 0·652 A, (b) 8·45 A
7. 0·1122 Ω, 280 W
8. 1·89 A
9. 187 Ω

Chapter 5

1. $L = 0.1$ H, $R = 20\ \Omega$
2. (a) 12·4 A, (b) 149 V, 311 V, 123 V, (c) 32 Hz
3. 3·94 A, 0·943 leading
4. (a) (i) 5·4 A, (ii) 3873 W. (b) (i) 3·1 A, (ii) 1290 W
5. (a) 72·2 kW, 85·75 kVA. (b) 122 A
6. 266 kVA, 370 A, 0·94 lagging
7. (a) 6·37 A. (b) 152·8 V across R, 320 V across L, and 123·3 V across C. (c) 32 Hz, 10 A
8. (a) I_R83 A, I_Y125 A, $I_B62.5$ A. (b) I_N73 A
9. (a) 5512 W, 9·58 A, 0·8 lagging
   (b) 16 536 W, 28·8 A, 0·8 lagging

Chapter 6

2. (a) 24 V, (b) 10·4 V
3. (a) 429 V, (b) 6·9 kW
4. (b) Heating reduced to 81 % of original value
5. D.C. if power factor less than 0·58
6. 424 V
7. 150 mm²
8. (b) 37 m, (c) 94 A

Chapter 7

7. (b) 394·8 V, (c) 2736 W

Chapter 9

2. 1167 rev/min, 1147 rev/min, 1126 rev/min
4. (a) 80·7 A, (b) 254 V, (c) 1·5 kW
6. 1422 rev/min
7. (b) 24·2 rev/s

Chapter 10

1. 152·5 A
2. 5·83 kW, 0·77 lagging, 43·2 %
8. (a) 0·27 m, (b) 4
9. (c) 81·3%

Chapter 11

1. (a) 96·5 %, (b) 95·8 %
2. (b) (i) 95·39 %, (ii) 94·84 %
4. (b) 104 kA, (c) 34·7 kA

Chapter 12

1. (*a*) 0·68 lagging, (*b*) 340 μF
4. (*b*) 0·84
5. (*a*) (*i*) £15 000, (*ii*) 1p, (*iii*) £1000
6. 515·3 μF

Chapter 14

4. (*a*) 0·0025 Ω, (*b*) 16661·7 Ω

Chapter 15

1. (*ii*) (*a*) 98, (*b*) 218
2. 61·7 lx, 82·48 lx
3. (*a*) 72, (*b*) 50
7. (*a*) (*i*) 70 lux, (*ii*) 40·6 lux
8. (*a*) (*i*) 25 lux, (*ii*) 18 lux
   (*b*) (*i*) 9·64 kW, (*ii*) 77 lamps

Chapter 16

3. (*b*) 23·2°C
5. 51·2 min
7. 16·5 kW

Chapter 19

7. (*b*) 9 days, (*c*) One day increase. B and E activities critical

INDEX